Selected Titles in This Series

746 **Peter Niemann,** Some generalized Kac-Moody algebras with known root multiplicities, 2002

745 **Mikhail A. Lifshits and Werner Linde,** Approximation and entropy numbers of Volterra operators with application to Brownian motion, 2002

744 **Roger Chalkley,** Basic global relative invariants for homogeneous linear differential equations, 2002

743 **Heng Sun,** Spectral decomposition of a covering of $GL(r)$: the Borel case, 2002

742 **J. E. Gilbert, Y. S. Han, J. A. Hogan, J. D. Lakey, D. Weiland, and G. Weiss,** Smooth molecular functions and singular integral operators, 2002

741 **Francisco Santos,** Triangulations of oriented matroids, 2002

740 **Rick Durrett,** Mutual invadability implies coexistence in spatial models, 2002

739 **Georgios K. Alexopoulos,** Sub-Laplacians with drift on Lie groups of polynomial volume growth, 2002

738 **Yasuro Gon,** Generalized Whittaker functions on $SU(2,2)$ with respect to the Siegel parabolic subgroup, 2002

737 **Arjen Doelman, Robert A. Gardner, and Tasso J. Kaper,** A stability index analysis of 1-D patterns of the Gray-Scott model, 2002

736 **Wojciech Chachólski and Jérôme Scherer,** Homotopy theory of diagrams, 2002

735 **Martina Brück, Xi Du, Joonsang Park, and Chuu-Lian Terng,** The submanifold geometries associated to Grassmannian systems, 2002

734 **Michel Van den Bergh,** Blowing up of non-commutative smooth surfaces, 2001

733 **Milé Krajčevski,** Tilings of the plane, hyperbolic groups and small cancellation conditions, 2001

732 **Jan O. Kleppe, Juan C. Migliore, Rosa Miró-Roig, Uwe Nagel, and Chris Peterson,** Gorenstein liaison, complete intersection liaison invariants and unobstructedness, 2001

731 **Jesús Bastero, Mario Milman, and Francisco J. Ruiz,** On the connection between weighted norm inequalities, commutators and real interpolation, 2001

730 **Suhyoung Choi,** The decomposition and classification of radiant affine 3-manifolds, 2001

729 **Michael Grosser, Eva Farkas, Michael Kunzinger, and Roland Steinbauer,** On the foundations of nonlinear generalized functions I and II, 2001

728 **Laura Smithies,** Equivariant analytic localization of group representations, 2001

727 **Anthony D. Blaom,** A geometric setting for Hamiltonian perturbation theory, 2001

726 **Victor L. Shapiro,** Singular quasilinearity and higher eigenvalues, 2001

725 **Jean-Pierre Rosay and Edgar Lee Stout,** Strong boundary values, analytic functionals, and nonlinear Paley-Wiener theory, 2001

724 **Lisa Carbone,** Non-uniform lattices on uniform trees, 2001

723 **Deborah M. King and John B. Strantzen,** Maximum entropy of cycles of even period, 2001

722 **Hernán Cendra, Jerrold E. Marsden, and Tudor S. Ratiu,** Lagrangian reduction by stages, 2001

721 **Ingrid C. Bauer,** Surfaces with $K^2 = 7$ and $p_g = 4$, 2001

720 **Palle E. T. Jorgensen,** Ruelle operators: Functions which are harmonic with respect to a transfer operator, 2001

719 **Steve Hofmann and John L. Lewis,** The Dirichlet problem for parabolic operators with singular drift terms, 2001

718 **Bernhard Lani-Wayda,** Wandering solutions of delay equations with sine-like feedback, 2001

717 **Ron Brown,** Frobenius groups and classical maximal orders, 2001

(Continued in the back of this publication)

Some Generalized Kac-Moody Algebras with Known Root Multiplicities

Number 746

Some Generalized Kac-Moody Algebras with Known Root Multiplicities

Peter Niemann

May 2002 • Volume 157 • Number 746 (second of 5 numbers) • ISSN 0065-9266

American Mathematical Society
Providence, Rhode Island

2000 *Mathematics Subject Classification.* Primary 17B65.

Library of Congress Cataloging-in-Publication Data

Niemann, Peter, 1965–
  Some generalized Kac-Moody algebras with known root multiplicities / Peter Niemann.
    p. cm. — (Memoirs of the American Mathematical Society, ISSN 0065-9266 ; no. 746)
  "Volume 157, number 746 (second of 5 numbers)."
  Includes bibliographical references.
  ISBN 0-8218-2888-6 (alk. paper)
  1. Kac-Moody algebras. 2. Root systems (Algebra)   I. Title.   II. Series.
QA3.A57 no. 746
[QA252.3]
510s—dc21
[512′.55]   2002018236

Memoirs of the American Mathematical Society

This journal is devoted entirely to research in pure and applied mathematics.

Subscription information. The 2002 subscription begins with volume 155 and consists of six mailings, each containing one or more numbers. Subscription prices for 2002 are $524 list, $419 institutional member. A late charge of 10% of the subscription price will be imposed on orders received from nonmembers after January 1 of the subscription year. Subscribers outside the United States and India must pay a postage surcharge of $31; subscribers in India must pay a postage surcharge of $43. Expedited delivery to destinations in North America $35; elsewhere $130. Each number may be ordered separately; *please specify number* when ordering an individual number. For prices and titles of recently released numbers, see the New Publications sections of the *Notices of the American Mathematical Society*.

Back number information. For back issues see the *AMS Catalog of Publications*.

Subscriptions and orders should be addressed to the American Mathematical Society, P. O. Box 845904, Boston, MA 02284-5904. *All orders must be accompanied by payment.* Other correspondence should be addressed to Box 6248, Providence, RI 02940-6248.

Copying and reprinting. Individual readers of this publication, and nonprofit libraries acting for them, are permitted to make fair use of the material, such as to copy a chapter for use in teaching or research. Permission is granted to quote brief passages from this publication in reviews, provided the customary acknowledgment of the source is given.

Republication, systematic copying, or multiple reproduction of any material in this publication is permitted only under license from the American Mathematical Society. Requests for such permission should be addressed to the Acquisitions Department, American Mathematical Society, P. O. Box 6248, Providence, Rhode Island 02940-6248. Requests can also be made by e-mail to reprint-permission@ams.org.

Memoirs of the American Mathematical Society is published bimonthly (each volume consisting usually of more than one number) by the American Mathematical Society at 201 Charles Street, Providence, RI 02904-2294. Periodicals postage paid at Providence, RI. Postmaster: Send address changes to Memoirs, American Mathematical Society, P. O. Box 6248, Providence, RI 02940-6248.

© 2002 by the American Mathematical Society. All rights reserved.
This publication is indexed in *Science Citation Index*®, *SciSearch*®, *Research Alert*®, *CompuMath Citation Index*®, *Current Contents*®/*Physical, Chemical & Earth Sciences*.
Printed in the United States of America.

∞ The paper used in this book is acid-free and falls within the guidelines established to ensure permanence and durability.
Visit the AMS home page at URL: http://www.ams.org/

10 9 8 7 6 5 4 3 2 1   07 06 05 04 03 02

Contents

Introduction	1
Chapter 1. Generalized Kac-Moody Algebras	**5**
1.1. Definition and Fundamental Properties	5
1.2. The Denominator Formula	9
1.3. Vertex Algebras	11
1.4. The Fake Monster Lie Algebra	15
1.5. The Twisted Denominator Formula	18
1.6. Construction of the GKMs	20
1.7. Root Multiplicities	24
Chapter 2. Modular Forms	**29**
2.1. Review of Modular Group and Modular Forms	29
2.2. Some Modular Forms Related to Eta	31
Chapter 3. Lattices and their Theta-Functions	**34**
3.1. Review of Results about Lattices	34
3.2. The Character of Theta	35
Chapter 4. The Proof of Theorem 1.7	**38**
4.1. The Theta-Function of L^*	38
4.2. Sums of Quadratic Residues	44
4.3. The Short Vectors of L^*	46
4.4. The Conclusion of the Proof	49
Chapter 5. The Real Simple Roots	**54**
5.1. The Set of Real Simple Roots	54
5.2. Holes and Dynkin Diagrams	56
5.3. The Volume Formula	63
5.4. The Automorphism Groups	67
Chapter 6. Hyperbolic Lie Algebras	**72**
6.1. Wan's classification	73
6.2. Finite, Affine, and Hyperbolic Subalgebras	73
6.3. Conclusions	89
Appendix A	**93**
N=23	93
N=11	94
N=7	94
N=5	95

    N=3   96
    N=2   99

Appendix B   110

Bibliography   116

Notation   118

ABSTRACT. Starting from Borcherds' fake monster Lie algebra we construct a sequence of six generalized Kac-Moody algebras whose denominator formulas, root systems and all root multiplicities can be described explicitly. The root systems decompose space into convex holes, of finite and affine type, similar to the situation in the case of the Leech lattice. As a corollary, we obtain strong upper bounds for the root multiplicities of a number of hyperbolic Lie algebras, including AE_3.

The author was supported by the Science and Engineering Research Council (UK), and Peterhouse, Cambridge.

Received by the editor August 16, 2000.

Introduction

In recent years the area of infinite-dimensional Lie algebras has attracted considerable attention because of its numerous connections with other topics in mathematics and, not least, its importance in theoretical physics. The state space of physical theories will sometimes be a representation space of an infinite-dimensional Kac-Moody algebra.

Surprisingly little is known about many obvious problems associated with such Lie algebras. One of these questions regards their root multiplicities, that is the dimensions of their root spaces. The cases of finite and affine Lie algebras are fully understood. Let us turn to more general Lie algebras of indefinite type which allow roots of negative norm. The problem now becomes far more complicated and only very partial answers are known. As [**Kac90**] remarks, the multiplicities of all roots of an indefinite-type Kac-Moody algebra are not known explicitly in any single case. At the same time, numerical calculations yield intriguing results, in particular for the simplest class of such Lie algebras which are called hyperbolic. They are defined by the condition on their Dynkin diagram that every subdiagram be of finite or affine type.

There are various kinds of results known about root multiplicities. There are global upper bounds for all Lie algebras which work well in some cases but are useless in others. There are recursive formulas which are useful for numerical calculations, notably the results of Kac and Peterson [**KP83**], and of Berman and Moody [**BM79**]. Furthermore, there are explicit results for some low level roots of selected algebras such as the treatments of Feingold and Frenkel [**FF83**] (on the algebra denoted AE_3 in the notation of [**Kac90**]), and Kac, Moody, and Wakimoto [**KMW88**] (on $E_{10} = T_{7,3,2}$). Recently, Gebert and Nicolai [**GN97**] produced some intriguing numerical results on the simple roots of E_{10} for level 2 and 3.

Borcherds introduces in [**Bor92**] the notion of generalized Kac-Moody algebras. These differ from ordinary Kac-Moody algebras in that they may possess imaginary simple roots. Generalized Kac-Moody algebras sometimes form the space of states for quantized chiral strings. It was a remarkable achievement that Borcherds then managed to construct a generalized Kac-Moody algebra, called the fake monster Lie algebra, and to give explicit root multiplicities for all its roots. The set of its roots can be identified with the 26-dimensional even unimodular lattice $II_{25,1} = \Lambda \oplus II_{1,1}$. Here, Λ stands for the 24-dimensional Leech lattice and $II_{1,1}$ denotes the unique even 2-dimensional unimodular Lorentzian lattice.

Borcherds suggested in [**Bor92**] that the fake monster Lie algebra might only be one of a whole class of generalized Kac-Moody algebras whose root multiplicities can be described explicitly. Let N be such that $N+1$ divides 24, let $M = \frac{24}{N+1}$. Thus N will be one of 2, 3, 5, 7, 11, or 23. Let σ be an automorphism of order N, cycle shape $1^M N^M$, of the Leech lattice. The aim of this thesis is to prove some of

Borcherds' conjectures by constructing a series of generalized Kac-Moody algebras $\mathcal{G}_N$ whose systems of simple roots can be identified with the fixed point lattices Λ^σ and certain elements of the dual lattices $\Lambda^{\sigma*}$. (Note that, unlike Λ, Λ^σ is no longer self-dual.) The Weyl denominator formula then becomes particularly simple and allows us to calculate the root multiplicities for these generalized Kac-Moody algebras explicitly.

We will now give a brief survey of the contents of the 6 chapters of this thesis. Chapter 1 recalls the basic definition and construction of the fake monster Lie algebra. Following [**Bor92**] we introduce the notion of twisted denominator formulas. They are obtained from the Weyl denominator formula of the fake monster Lie algebra by the action of the Leech lattice automorphism σ. We then proceed to formulate the main theorem of this work which claims the existence of a series of generalized Kac-Moody algebras and states their root multiplicities explicitly. Following ideas of [**Bor92**] the proof is reduced to an equality between two modular forms. One of these is the θ-function of the lattice $\Lambda^{\sigma*}$. The other function is derived from the Dedekind η-function.

Both the above functions are modular forms with respect to some subgroup of the modular group, which has finite index. The strategy of the proof must now be as follows. We check that the modularity properties of both functions are equal and that a sufficient number of leading coefficients in a Laurant expansion around zero coincide. As the space of modular forms with respect to such a modular subgroup is finite-dimensional this shows the claimed equality. It is easy to see that, the larger the transformation group is, the smaller will be the dimension of the space of modular forms, thus requiring fewer leading coefficients in the remainder of the argument. The most suitable group for our purposes turns out to be $\Gamma_0(N)$.

Chapter 2 recalls the basic notions of modular group and modular form, as well as the Dedekind η-function. We then proceed to define the specific modular forms we need in this work and establish their precise modularity properties under the elements of $\Gamma_0(N)$.

Chapter 3 recalls some basic results about the lattices in question and then determines the exact transformation properties, including characters, of their θ-functions under the elements of $\Gamma_0(N)$.

In chapter 4 we determine the leading coefficients of the above modular forms. This is straightforward for those modular forms which were derived from the η-function. For θ, however, this corresponds to the enumeration of the short vectors of $\Lambda^{\sigma*}$. For $N = 2$ and $N = 3$ it is well known that the fixed point lattices are the Barnes-Wall lattice Λ_{16} and the Coxeter-Todd lattice K_{12}. They have been considered in the literature in their own right. For the remaining N we devise a strategy to count the short vectors. This makes use of the fact that the lattices in question are all induced from the Leech lattice which in turn is built upon the 24-dimensional Golay code. It is possible to reduce the required enumeration in $\Lambda^{\sigma*}$ to one of Golay code elements.

We then put everything together and obtain the desired equality, thus completing the proof of the main theorem of this work. We thus have proven the existence of a series of generalized Kac-Moody algebras $\mathcal{G}_N$ and have determined their root multiplicities explicitly. At the same time, their Weyl denominator formulas can be interpreted as new combinatorial identities, similar to the Macdonald identities.

In the second part of this work we investigate the generalized Kac-Moody algebras $\mathcal{G}_N$ in more detail. In chapter 5 we determine their simple roots. We then identify this set with a set $\mathcal{R}$ consisting of the elements of the fixed point lattice Λ^σ and some elements of its dual lattice $\Lambda^{\sigma*}$. It is well known (see e.g. [**CS88**], chapter 25) that the elements of the Leech lattice generate a decomposition of space into convex holes of radius less or equal to $\sqrt{2}$. Now the elements of the Leech lattice can be identified with the simple roots of the fake monster Lie algebra. It turns out that holes of radius $\sqrt{2}$ correspond to affine subalgebras, whereas holes of radius less than $\sqrt{2}$ correspond to subalgebras of finite type. We generalize this to the sets $\mathcal{R}$. The main difference is that now the algebras have simple roots of two different lengths. We therefore have to generalize the notion of radius and centre accordingly. If we do so it remains true that the decomposition produces holes of (generalized) radius less or equal $\sqrt{2}$, corresponding to subalgebras of finite and affine type respectively. We further show that all subalgebras of $\mathcal{G}_N$ of finite or affine type can be identified as (generalized) holes in $\mathcal{R}$. As a corollary, we can relate the covering radius of the fixed point lattices Λ^σ to the covering radius of the Leech lattice.

One of our aims in chapter 6 is the complete classification of all (generalized) holes of $\mathcal{R}$. The remainder of chapter 5 therefore develops a number of techniques which will enable us to carry out the decomposition and check its correctness. One such check is the volume formula which simply states that the sum of the volumes of the individual holes must be the total volume of space. We therefore determine the volumes of any finite and affine holes. Another test concerns the automorphism groups of the individual holes, related to the automorphism group of the fixed point lattice as a whole. We derive a number of technical results, relating the generalized holes of $\mathcal{R}$ to the (known) holes in the decomposition of the Leech lattice.

Chapter 6 carries out the classification. This presents no theoretical problems but involves long and repetitive calculations which may only be carried out by computer. The main purpose of the first two sections of chapter 6 is to demonstrate how the various results of chapter 5 come together to achieve the classification and why the output of a computer program constitutes a mathematical proof. To this end, we give the explicit calculations of the two simplest cases, that is $N = 23$ and $N = 11$. The case $N = 23$ is almost trivial but very useful as it is 2-dimensional and so helps to visualize the problem. The 4-dimensional case $N = 11$ is already sufficiently general to demonstrate how the program acts. We can therefore restrict ourselves to giving the input data for the remaining cases. The complete classification is given as appendix A. For the most complicated case $N = 2$ there are 475 different types of (generalized) holes in $\mathcal{R}$.

As a corollary to this decomposition we can now identify all hyperbolic subalgebras of the $\mathcal{G}_N$. The known root multiplicities of $\mathcal{G}_N$ then form upper bounds for the root multiplicities of these hyperbolic Lie algebras. There are many interesting examples for which the upper bounds obtained by the above technique improve on the upper bounds given in the existing literature. Of particular interest is the hyperbolic Lie algebra AE_3 which was also investigated in [**FF83**]. The algebra AE_3 is defined by the following Cartan matrix:

$$\begin{pmatrix} 2 & -2 & 0 \\ -2 & 2 & -1 \\ 0 & -1 & 2 \end{pmatrix}.$$

We obtain strong upper bounds for the root multiplicities of this algebra. These bounds are close to some existing conjectures of [**Kac90**] (see exercise 13.37). Our results explain why the multiplicities of the roots in the algebra AE_3 are often equal to the values of a partition function $p_n(1 - r^2/2)$.

However, for other hyperbolic Lie algebras our new upper bounds are not always sharp or at least an improvement on existing ones. In section 6.2 we list the successful cases. In appendix B we provide some numerical data for all hyperbolic Lie algebras of rank 7 to 10. The results for the algebra $T_{4,3,3}$ are of particular interest as they show that there are roots r of multiplicities both larger and smaller than the partition function $p_6(1 - r^2/2)$. In the concluding section 6.3 we derive some conditions which are necessary if we want to create useful upper bounds. This will then enable us to conjecture how far the strategy of this work may be extended to other generalized Kac-Moody algebras.

Acknowledgements: I wish to thank Richard Borcherds, my research supervisor, for suggesting many of the problems discussed in this work and for the advice and encouragement he provided throughout my research. Furthermore, I would like to thank Simon Norton for his patient explanations of the ATLAS, and Elizabeth Jurisich for clarifications regarding the nature of specializations. Finally, I would like to thank the referee for valuable comments on the first version of this paper.

CHAPTER 1

Generalized Kac-Moody Algebras

The main object of study in this work is a series of generalized Kac-Moody algebras whose root multiplicities will be determined explicitly. This chapter introduces the concept and describes the construction of these generalized Kac-Moody algebras. Chapters 2-4 will then complete the proof of their existence and their root multiplicity formulas. It must be emphasised in this context that chapters 2-4 are independent of the results of the present chapter 1. Chapter 1 has been placed in its position ahead of chapters 2-4 only to provide the framework identifying which auxiliary calculations are required in chapters 2-4. Sections 1.1 and 1.2 recall the definition and elementary properties of generalized Kac-Moody algebras, and specifically their Weyl denominator formula. The series of generalized Kac-Moody algebras in this work will be derived from the fake monster Lie algebra as introduced in [**Bor90b**]. This, in turn, is constructed from subspaces of a certain vertex algebra. In section 1.3 we therefore briefly recall the construction of the vertex algebra in question. Section 1.4 outlines the construction and some properties of the fake monster Lie algebra, following [**Bor92**]. Its root lattice can be identified with the Leech lattice and all root multiplicities are known explicitly. Section 1.5 quotes [**Bor92**] to recall how automorphisms of the Leech lattice can be applied to the denominator formula of the fake monster Lie algebra, resulting in new identities which may be regarded as 'twisted' denominator formulas. In sections 1.6 and 1.7 we restrict our attention to a specific series of six Leech lattice automorphisms. For these we show how the new 'twisted' identities can be interpreted as denominator formulas of six new generalized Kac-Moody algebras whose root multiplicities can again be calculated explicitly. For general automorphisms, the new algebras will not necessarily be generalized Kac-Moody algebras but may be Lie superalgebras where negative root multiplicities may occur. Section 1.7, in conjunction with chapters 2-4, establishes an explicit form of the denominator formulas (equation 1.26), for the series of six new generalized Kac-Moody algebras, corresponding to the series of Leech lattice automorphisms. This yields an explicit (non-recursive) formula for the root multiplicities of the generalized Kac-Moody algebras (corollary to theorem 1.7), and carries out a programme suggested in [**Bor92**] (14. Examples 1 and 2).

1.1. Definition and Fundamental Properties

This section establishes the definition of generalized Kac-Moody algebras which we will use within this work. Numerous variants have been proposed in recent publications. We will follow the approach of [**Jur98**], but specialize to the variant considered in [**Bor92**] when appropriate. Throughout this work we will use the initials GKM for generalized Kac-Moody algebra.

1.1.1. The Definition. [**Jur98**] defines GKMs through generalized Cartan matrices. A real matrix $C = (c_{ij})$, i, j in some index set I (possibly countably infinite), shall be called a generalized Cartan matrix if it satisfies the following conditions:

(C1) C is symmetric;
(C2) $c_{ij} \leq 0$ if $i \neq j$;
(C3) if $c_{ii} > 0$ then $2c_{ij}/c_{ii} \in \mathbb{Z}$ for all $j \in I$.

[**Jur98**] defines the generalized Kac-Moody algebra (GKM) $G = G(C)$ associated with this generalized Cartan matrix to be the Lie algebra generated by elements e_i, f_i, h_i, for $i \in I$, where the generators satisfy the following relations for $i, j, k \in I$:

(R1) $[h_i, h_j] = 0$
(R2) $[e_i, f_j] = \delta_i^j h_i$
(R3) $[h_i, e_k] = c_{ik} e_k, [h_i, f_k] = -c_{ik} f_k$
(R4) If $c_{ii} > 0$ and $i \neq j$ then $\text{ad}(e_i)^n e_j = \text{ad}(f_i)^n f_j = 0$, where $n = 1 - 2c_{ij}/c_{ii}$.
(R5) If $c_{ii} \leq 0$, $c_{jj} \leq 0$, and $c_{ij} = 0$ then $[e_i, e_j] = [f_i, f_j] = 0$.

We follow [**Jur98**] in restricting the definition to symmetric Cartan matrices as we will not encounter any non-symmetrisable Cartan matrices in the context of the present work.

As in the finite dimensional case, the elements h_i, $i \in I$, span an abelian subalgebra H of $G(C)$, called its Cartan subalgebra. Let E be the subalgebra generated by the e_i, $i \in I$, and F be the subalgebra generated by the f_i, $i \in I$. Then the GKM $G(C)$ has the triangular decomposition [**Jur96**]

(1.1) $$G(C) = E \oplus H \oplus F.$$

Every non-zero ideal of $G(C)$ has non-zero intersection with H. The centre of $G(C)$ is contained in H. [**Jur96**]

1.1.2. Roots, Central Ideals, Central Extensions. Roots of finite dimensional Lie algebras are commonly defined as elements of the dual space of the Cartan subalgebra. Hence, quotienting out some ideal of the Cartan subalgebra, or extending the Cartan subalgebra centrally will also affect the space where roots are defined. For the definition of roots for GKMs, there consequently exist various options which relate to the choice of central extension for the GKM defined above.

Alternatively, the approach of [**Bor92**] defines roots as a free abelian group (of abstract symbols), and identifies a natural homomorphism to the elements h_i of the Cartan subalgebra H.

If we consider roots within the dual space H^* with $G(C)$ constructed as above, the simple roots will not necessarily be linearly independent. [**Jur96**] therefore extends the GKM $G(C)$ by an algebra of 'degree derivations'. This increases the dimension of the Cartan subalgebra and ensures that the simple roots will be defined linearly independent. Note that, besides the approaches of [**Jur96**] and [**Bor92**], others have been proposed, such as the approach described in [**Kac90**] whose 'realization' of the Cartan matrix again guarantees the linear independence of the resulting simple roots.

As is discussed in detail in [**Jur98**], the denominator formula for an arbitrary Cartan matrix with linearly dependent simple roots will not necessarily be well defined as some terms may be infinite. A general theory can therefore only be formulated in a framework which guarantees the linear independence of the simple roots. This will either be achieved as in [**Bor92**], or through a suitable extension of the Cartan subalgebra as in [**Jur96**]. Nevertheless, it may be possible to define denominator formulas for certain other Lie algebras, through a process called 'specialization' in [**Jur96**]. Such specializations will quotient out subalgebras of the (extended) Cartan subalgebra. This operation will also quotient out the corresponding subspaces of the dual space where roots are defined. As is pointed out in the remark following definition 3 of [**Jur98**], specializations are valid, as long as they are well defined. For the GKMs constructed in the present work we will find that the denominator formula remains well defined under specialization.

Where simple roots are linearly independent they obviously have multiplicity one. The process of specialization may map multiple, distinct simple roots to the same image. Therefore, following specialization, some simple roots may have multiplicity greater than one.

The approach of Jurisich. [**Jur96**] extends the GKM by all 'degree derivations'. Let $\deg(e_i) = -\deg(f_i) = (0, \ldots, 0, 1, 0, \ldots)$ where 1 appears in the ith position, and let $\deg(h_i) = (0, \ldots)$. Degree derivations d_i are defined by letting d_i act on the degree $(n_1, n_2, \ldots)$ subspace of the GKM as multiplication by the scalar n_i. [**Jur96**] defines the extended Lie algebra $G^e = G^e(C)$ as the semidirect product of the GKM $G(C)$ with the space of all degree derivations. The extension is central. Let H^e denote the Cartan subalgebra of $G^e(C)$. The extended Lie algebra has the decomposition

$$(1.2) \qquad G^e(C) = E \oplus H^e \oplus F.$$

Roots of $G^e(C)$ defined in the space H^{e*} are necessarily linearly independent.

The extended Lie algebra $G^e(C)$ is clearly not identical to the GKM $G(C)$ constructed from the original Cartan matrix C, but may be considered naturally associated with it. More generally, if a Lie algebra can be mapped to $G(C)$ modulo some central ideals, some central extensions, or outer derivations (as is the case in specializations), we will consider that Lie algebra naturally associated with the Cartan matrix C and the GKM $G(C)$.

Having selected the Cartan algebra H^e, roots may be defined as in the finite dimensional case. For $r \in H^{e*}$, let

$$G^r = \{ \, x \in G \mid [h, x] = r(h)x \text{ for all } h \in H^e \}.$$

The roots of G are the non-zero elements r of H^{e*} such that $G^r \neq 0$. The elements $r_i \in H^{e*}$ such that the generators e_i are in G^{r_i} are called simple roots. G^r is the root space of $r \in H^{e*}$. A root r is called positive if it is the sum of simple roots, and negative otherwise.

The approach of Borcherds. [**Bor92**] defines the root lattice of $G(C)$ as the free abelian group generated by elements r_i, for $i \in I$, with the bilinear form given by $(r_i, r_j) = c_{ij}$. Here, the c_{ij} are the elements of the symmetric generalized Cartan matrix C. The elements r_i correspond to the simple roots. The GKM is graded by the root lattice if we let e_i have degree r_i and f_i have degree $-r_i$. If r is in the root lattice then the vector space of elements of the Lie algebra of that degree is called the root space of r. There is a natural homomorphism of abelian groups from the root lattice to the Cartan subalgebra H taking r_i to h_i which preserves the bilinear forms. This homomorphism will not necessarily be injective. If, for example, the null space of the bilinear form has been quotiented out, this may have introduced relations among the elements h_i of H.

1.1.3. Norm, Weyl Vector, Weyl Chamber and Cartan Involution.
We define the norm of an element r of the root lattice as the scalar product

$$\text{norm}(r) = (r, r). \tag{1.3}$$

Note that we use the square of the standard norm, because the scalar product is not positive definite. A root of a GKM is called real if it has positive norm (r,r), and imaginary otherwise.

In the framework of the extended root space H^{e*}, the Weyl vector ρ is any element of H^{e*} which satisfies $(\rho, r_i) = -(r_i, r_i)/2$. Note that this implies $\rho(h_i) = -c_{ii}/2$. The (fundamental) Weyl chamber is the set of vectors v of the root space H^{e*} which satisfy $(v, r_i) \leq 0$ for all real simple roots r_i. Equivalently, [**Bor92**] defines the Weyl vector as the additive map from the free abelian group of roots to $\mathbb{R}$, taking r_i to $-(r_i, r_i)/2$ for all $i \in I$. The (fundamental) Weyl chamber is the set of all vectors $v \in H$ with $(v, h_i) \leq 0$ for all $h_i \in H$ that correspond to real simple roots. Following [**Bor90b**], we define the height of a root r as $-(\rho, r)$.

Note that [**Bor92**] uses non-standard sign conventions in this place, and the present paper will follow his conventions. In the case of a specialization, where simple roots may be linearly dependent, there is no reason why a Weyl vector should exist in general. In the example of the affine algebra A_2, the relation $r_2 = -r_1$ of the two simple roots (which holds in unextended 2-dimensional dual space) shows that no Weyl vector can exist.

$G(C)$ has an involution ω with $\omega(e_i) = -f_i$, $\omega(f_i) = -e_i$, called the Cartan involution. There is a unique invariant bilinear form $(\cdot, \cdot)$ on $G(C)$ such that $(e_i, f_i) = 1$ for all i, and it also has the property that $-(g, \omega(g)) > 0$ whenever g is a homogeneous element of non-zero degree.

1.1.4. Universal Central Extension.
Let C be a generalized Cartan matrix, satisfying conditions (C1) to (C3). For an alternative characterisation of some GKMs, [**Bor92**] defines the universal generalized Kac-Moody algebra (UGKM) $U(C)$ of this matrix to be the Lie algebra generated by elements e_i, f_i, h_{ij}, for $i, j \in I$ satisfying the following relations $(i, j, k, l \in I)$:

(U1) $[h_{ij}, h_{kl}] = 0$
(U2) $[e_i, f_j] = h_{ij}$
(U3) $[h_{ij}, e_k] = \delta_i^j c_{ik} e_k$, $[h_{ij}, f_k] = -\delta_i^j c_{ik} f_k$
(U4) If $c_{ii} > 0$ and $i \neq j$ then $\text{ad}(e_i)^n e_j = \text{ad}(f_i)^n f_j = 0$, where $n = 1 - 2c_{ij}/c_{ii}$.
(U5) If $c_{ii} \leq 0$, $c_{jj} \leq 0$, and $c_{ij} = 0$ then $[e_i, e_j] = [f_i, f_j] = 0$.

The UGKM is therefore an extension of the GKM defined in section 1.1.1 above. It is extended precisely by the additional central generators $h_{ij}, i \neq j$. [**Bor92**] observes that the element h_{ij} is 0 unless the i'th and j'th column of the Cartan matrix C are equal. The elements h_{ij} for which the i'th and j'th column are equal form a basis of the Cartan subalgebra H of $U(C)$. In the case of ordinary Kac-Moody algebras, the i'th and j'th column of C cannot be equal unless $i = j$, so the only non-zero elements h_{ij} are those of the form h_{ii}, which are usually denoted by h_i. The centre of $U(C)$ contains all elements h_{ij} for $i \neq j$.

Using the definition of the universal extension, [**Bor92**] provides an alternative characterisation of some GKMs (see also [**Jur98**]).

THEOREM 1.1. *Suppose that G is a Lie algebra satisfying the following three properties:*

(1) *G can be $\mathbb{Z}$-graded as $G = \bigoplus_{i \in \mathbb{Z}} G_i$, and G_i is finite dimensional if $i \neq 0$.*

(2) *G has an involution ω which maps G_i into G_{-i} and acts as -1 on G_0.*

(3) *G has a Lie algebra invariant bilinear form $(\cdot, \cdot)$, which is also invariant under ω such that G_i and G_j are orthogonal if $i \neq -j$, and such that $-(g, \omega(g)) > 0$ if g is a nonzero homogeneous element of G of nonzero degree.*

Then there is a unique UGKM, graded by putting $\deg(e_i) = -\deg(f_i) = n_i$ for some positive integers n_i, with a homomorphism f (not necessarily unique) to G such that

(a) *f preserves the gradings, involutions and bilinear forms (as defined above).*

(b) *The kernel of f is in the centre of the UGKM (which is contained in the abelian subalgebra spanned by the elements h_{ij}).*

(c) *The image of f is an ideal of G, and G is the semidirect product of this subalgebra and a subalgebra of the abelian subalgebra G_0. Moreover, the images of all the generators e_i and f_i are eigenvectors of G_0.*

[**Bor92**] defines GKMs through properties (1) to (3) of theorem 1.1. [**Jur98**] points out that the converse of theorem 1.1 is not true: GKMs constructed from generalized Cartan matrices cannot necessarily be graded satisfying both conditions (1) and (3). For this reason, the present work follows [**Jur98**] and adopts the wider definition of GKMs directly from generalized Cartan matrices. All GKMs considered in this work do, however, permit a grading of the type described in theorem 1.1.

The above exposition shows that the only major difference between GKMs and ordinary Kac-Moody algebras is that GKMs may have imaginary simple roots. A further generalization of GKMs are Lie superalgebras. Here, we allow the imaginary simple roots to have negative multiplicity. These are then called superroots. The Cartan matrix of a Lie superalgebra may depend on the $\mathbb{Z}$-grading chosen. We will not encounter Lie superalgebras in the course of this work.

1.2. The Denominator Formula

We recall from equations (1.1) and (1.2) that any GKM can be written as the direct sum $E \oplus H \oplus F$ where H is the Cartan subalgebra and E and F are the subalgebras corresponding to the positive and negative roots. The homology groups of a Lie algebra are defined as the homology groups of the standard sequence of exterior powers,

$$(1.4) \qquad \cdots \to \bigwedge\nolimits^2(E) \to \bigwedge\nolimits^1(E) \to \bigwedge\nolimits^0(E) \to 0,$$

(see Cartan and Eilenberg's introduction to Lie algebra homology, [**CE56**].) We consider the following two virtual vector spaces.

$$\bigwedge(E) = \bigwedge\nolimits^0(E) \ominus \bigwedge\nolimits^1(E) \oplus \bigwedge\nolimits^2(E) \ldots,$$

which is the alternating sum of the exterior powers of E, and

$$H_*(E) = H_0(E) \ominus H_1(E) \oplus H_2(E) \ldots,$$

which is the alternating sum of the homology groups $H_i(E)$. If L is the root lattice of the Lie algebra then both spaces are L-graded virtual vector spaces whose homogeneous pieces are finite dimensional, so the infinite sums are meaningful. From the definitions it follows that

$$\bigwedge(E) = H_*(E), \tag{1.5}$$

as virtual L-graded vector spaces (Euler-Poincare principle). This formula can be used as a starting point to calculate the denominator formula for the GKM. To do this, [**Bor92**] identifies the spaces $H_i(E)$ and then calculates the formal character

$$\chi(V) = \sum_{\lambda \in L} (\dim V_\lambda) e^\lambda, \tag{1.6}$$

on both sides. The left hand term of (1.5), $\bigwedge(E)$, can be dealt with by a standard combinatorial argument, counting occurrences. The argument is analogue to that in the case of ordinary Kac-Moody algebras, which is well documented, see, for example, chapter 10 of [**Kac90**]. Let us consider the right hand term of (1.5), $H_*(E)$. [**Bor92**] reports that the techniques developed by Garland and Lepowsky in [**GL76**] can be adapted. A detailed discussion can be found in [**Jur96**] (theorem 3.13).

THEOREM 1.2. *Let G be a GKM with Weyl group W, Weyl vector ρ, and root lattice L.*

a) $H_i(E)$ is the subspace of $\bigwedge^i(E)$ spanned by the homogeneous vectors of $\bigwedge^i(E)$ whose degrees $r \in L$ satisfy $(r + \rho)^2 = \rho^2$.

b) Let S denote the subspace of $H(E)$ of elements whose degree r has the property that $r + \rho$ is in the fundamental Weyl chamber. Then S is isomorphic to the subspace of $\bigwedge(E)$ of all elements that can be written in the form $e_1 \wedge e_2 \wedge \ldots$ where the e_i's are vectors in the root spaces of pairwise orthogonal imaginary roots.

It must be recalled in this context that the fundamental Weyl chamber for GKMs is still determined by the *real* simple roots. Using theorem 1.2, [**Bor92**] calculates the formal character of $H_*(E)$. Again, a detailed discussion can be found in [**Jur96**] (theorem 3.16).

THEOREM 1.3. *Let G be a GKM with Weyl group W, Weyl vector ρ, root lattice L, and denote the positive roots by L^+. If $w \in W$ then $\det(w)$ is defined to be $+1$ or -1, depending on whether w is the product of an even or odd number of reflections. (If the root lattice is finite dimensional this is just the usual determinant of w.) We define $\epsilon(\alpha)$ for $\alpha \in L$ to be $(-1)^n$ if α is the sum of a set of n pairwise orthogonal imaginary simple roots, and 0 otherwise.*

Then the denominator formula of G is

$$e^\rho \prod_{\alpha \in L^+} (1 - e^\alpha)^{\text{mult}(\alpha)} = \sum_{w \in W} \det(w) w \left(e^\rho \sum_{\alpha \in L^+} \epsilon(\alpha) e^\alpha \right).$$

Thus, the formula is very similar to the well known denominator formula for ordinary Kac-Moody algebras and reduces to it if there are no imaginary simple roots. In fact, the sum over $\alpha \in L^+$ is precisely the character of the subspace S described in theorem 1.2b. If there are no imaginary simple roots this sum collapses to 1. Note that the definition of $\epsilon(\alpha)$ in theorem 1.3 assumes that the simple roots are linearly independent. In the case of a specialization, $\epsilon(\alpha)$ will denote the sum over all relevant terms $(-1)^{n_i}$, each corresponding to a representation of α as the sum of n_i simple roots.

We can recover the full character formula for GKMs in the same way, starting, in the place of (1.4), with the generalized chain complex whose vector spaces are spaces $\bigwedge^j(E,V)$, that is spaces $\bigwedge^j(E)$, tensored with any lowest weight Lie algebra module V. We will, however, not make use of this in the remainder of this paper.

For more information on the significance of especially the term related to imaginary simple roots, we refer to the detailed discussion in [**Jur96**]. The individual terms of the denominator formula will depend on the selection of the Cartan subalgebra, and thus the root space. The identity is valid as an identity in the free abelian group, as in the approach of [**Bor92**]. Equally, the formula is meaningful and valid in general if, as in the approach of [**Jur96**], we work with a suitably extended Lie algebra G^e in the place of the GKM $G(C)$ so that all simple roots are linearly independent.

As we have seen, the roots of the unextended GKM $G(C)$ defined in the space H^* will not necessarily be linearly independent. [**Jur98**] discusses that this may lead to denominator identities with infinite terms and other problems. The affine algebra A_2 provides a simple example of such problems: Here, the two real simple roots satisfy $r_2 = -r_1$ in the unextended root space.

There are, however, circumstances where it is possible to formulate results in H^* through specialization of the results in the extended root space H^{e*}. This specialization is achieved through projection from H^{e*} to the space H^*. As an abstract identity, the specialization of the denominator formula remains valid provided no multiplicities become infinite. Note that we may encounter further problems related to the interpretation of that abstract identity as a denominator formula. Examples of potential ambiguities are the identification of simple and non-simple roots, or of positive and negative roots. [**Jur98**] discusses why specialization works for Borcherds' monster Lie algebra. Similarly, we will find that this is also the case for all GKMs constructed in this present work.

1.3. Vertex Algebras

Vertex algebras had already been used extensively in theoretical physics, when Borcherds [**Bor86**] formalized the definition and showed how to construct vertex algebras for any even lattice. We follow [**Bor92**] to give a brief survey of the definitions and results which will be relevant to the present work.

A vertex algebra over the real numbers is a vector space V over $\mathbb{R}$ with an infinite number of bilinear products, written $u_n v$, where $u, v, u_n v \in V$ and $n \in \mathbb{Z}$, such that

(1) $u_n v = 0$ for n sufficiently large (depending on u and v).
(2)
$$\sum_{i \in \mathbb{Z}} \binom{m}{i} (u_{q+i}v)_{m+n-i} w$$
$$= \sum_{i \in \mathbb{Z}} (-1)^i \binom{q}{i} \left(u_{m+q-i}(v_{n+i}w) - (-1)^q (v_{n+q-i}(u_{m+i}w)) \right)$$

for all u, v, and w in V and all integers m, n, and q.
(3) There is an element $1 \in V$ such that $v_n 1 = 0$ if $n \geq 0$ and $v_{-1} 1 = v$.

The operators u_n may then be combined into the vertex operator
$$Q(u,z) = \sum_{n\in\mathbb{Z}} u_n z^{-n-1}$$
which is an operator valued formal Laurent series in the formal variable z.

[**Bor86**] defines an operator D on the vertex algebra by $D(v) = v_{-2}1$. The vector space V/DV is a Lie algebra, where the bracket is defined by $[u,v] = u_0 v$.

A conformal vector of dimension or central charge $c \in \mathbb{R}$ of a vertex algebra V is defined to be an element ω of V such that
(1) $\omega_0 v = D(v)$ for any $v \in V$,
(2) $\omega_1 \omega = 2\omega$,
(3) $\omega_3 \omega = c/2$,
(4) $\omega_i \omega = 0$, if $i = 2$ or $i > 3$,
(5) any element of V is a sum of eigenvectors of ω_1 with integral eigenvalues.

[**Bor86**] defines the operators L_i on V for $i \in \mathbb{Z}$ by $L_i = \omega_{i+1}$. It can then be shown that the operators L_i satisfy the relations
$$[L_i, L_j] = (i-j)L_{i+j} + \binom{i+1}{3}\frac{c}{2}\delta^i_{-j}.$$

They make V into a module over the Virasoro algebra. For $n \in \mathbb{Z}$, the physical space P^n is defined to be the space of vectors $w \in V$ such that

(1.7a) $$L_0(w) = \omega_1(w) = nw$$
(1.7b) $$L_i(w) = 0, \text{ if } i > 0.$$

The space $P^1/(DV \cap P^1)$ is a subalgebra of the Lie algebra V/DV. In the cases considered in [**Bor92**] (and thus in the cases considered in this work) this is equal to P^1/DP^0.

In [**Bor86**] examples of vertex algebras are constructed from lattices. We present the explicit spaces, operators, and elements, both as an illustration of the above definitions and for later reference. The results are well-known but not necessarily in the context of Borcherds' framework. Let L be an even lattice. There exists a central extension by a group of order 2, $\hat{L}$, which is uniquely characterized by the following properties. The elements of $\hat{L}$ will be written $\epsilon^n e^r$, $r \in L$, where $n = 0$ or $n = 1$. The commutator is

(1.8) $$e^{r_1} e^{r_2} = \epsilon^{(r_1, r_2)} e^{r_2} e^{r_1} \quad \text{where} \quad \epsilon^2 = 1.$$

The underlying vector space $V(L)$ of the vertex algebra associated with L is defined as follows;

(1.9) $$V(L) = \mathbb{R}(\hat{L}) \bigotimes S(\bigoplus_{i>0}(L_{(i)} \otimes \mathbb{R})).$$

In [**Bor86**], this space is referred to as Fock space. This term has since been used to denote a slightly different space so that we will not use the term. $\mathbb{R}(\hat{L})$ is the twisted group ring of $\hat{L}$ and $S = S(\bigoplus_{i>0}(L_{(i)} \otimes \mathbb{R}))$ is the symmetric algebra on the sum of a countable number of copies of the lattice L. Thus, a general element of $V(L)$ will be a linear combination of elements of the form

(1.10) $$v = e^r \prod_{i=1}^q t_i(p_i)$$

where $r, t_i \in L$, $q \geq 0$, $p_i \geq 1$, and $t_i(p_i)$ is an element of the copy $L_{(p_i)} \otimes \mathbb{R}$ within the symmetric algebra, and the integers p_i are not necessarily distinct.

In order to construct vertex operators for all elements, we begin by defining 'annihilation' and 'creation' operators. For $t \in L$, and $j \in \mathbb{Z}$, define $t(j)$ as a linear map on $V(L)$. It is fully characterized by its action on elements of $V(L)$ of the form (1.10).

if $j > 0$ then $t(j)$ is multiplication by $t(j)$;
if $j = 0$ then $t(0)v = (t,r)v$;
if $j < 0$ then $t(j)$ acts as a derivation so that $t(j)e^r = 0$, and
$$t(j)t_i(p_i) = -j(t, t_i)\delta_{p_i}^{-j}.$$

We define the vertex operator of a general element of $V(L)$ in three steps. Let $t \in L$, $p > 0$, let $z \in \mathbb{C}$ be a complex number, and let v be as in (1.10).

$$Q(t, z) = \sum_{j \neq 0} t(j) \frac{z^j}{j} + t(0)\log z + t$$

$$Q(t(p), z) = \frac{1}{(p-1)!}\left(\frac{d}{dz}\right) Q(t, z)$$

$$Q(v, z) = Q\left(e^r \prod t_i(p_i), z\right) = \; :e^{Q(r,z)} \prod Q(t_i(p_i), z):$$

Here the ':' is the standard notation for normal ordering, such that all 'creation' operators (e^r, $t(p), p \geq 1$) occur to the left of all 'annihilation' operators ($t(p), p \leq 0$). The product $v_n(w)$ for $n \in \mathbb{Z}$ and $w \in V(L)$ is then defined as the coefficient of z^{-n-1} in $:Q(v,z):(w)$.

Choose a basis $s_i, i \in I$, of the lattice L, and a dual basis s'_i. Then the vector

$$\omega = \frac{1}{2}\sum_{i \in I} s_i(1)s'_i(1)$$

is a conformal vector in $V(L)$. Its central charge can be identified as the dimension of the lattice L. Applying the general vertex operator construction to ω, we identify the operators $L_n = \omega_{n+1}$ as

$$L_n = \sum_{j \in \mathbb{Z}} \sum_{i \in I} :s_i(j)s'_i(-n-j):$$

In particular, we note the action of L_0 on homogeneous elements of the form (1.10)

(1.11) $$L_0 v = \left(\frac{r^2}{2} + \sum_i p_i\right)v.$$

Thus, L_0 defines a $\mathbb{Z}$-grading which we will refer to below as $\deg_\mathbb{Z}$, or $\deg_\mathbb{Z}^{(L)}$ if we require to specify the referenced underlying lattice L. L_{-1} is the derivation D. For $r, t \in L$, and $p \geq 1$ we obtain $L_{-1}e^r = r(1)e^r$, and $L_{-1}t(p) = pt(p+1)$.

Suppose A is an ordinary Kac-Moody algebra with simple roots a_i of norm 2. The derived algebra A' is the algebra generated by generators e_i, f_i, h_i for each simple root a_i. If the lattice L contains the root lattice of A then we can map the derived algebra A' of A to P^1/DP^0, as follows; $e_i \mapsto e^{a_i}$, $f_i \mapsto e^{-a_i}$, $h_i \mapsto a_i(1)$.

As an example of the definitions and constructions above, we calculate the bracket $[e_i, h_j] = (e_i)_0(h_j)$ in P^1/DP^0.

$$Q\left(e^{a_i}, z\right) a_j(1) = e^{a_i} \exp\left(\sum_{n>0} \frac{a_i(n)}{n} z^n\right) \exp\left(\sum_{n<0} \frac{a_i(n)}{n} z^n\right) z^{a_i(0)} a_j(1)$$

Here, $z^{a_i(0)}$ acts on $e^r \in V(L)$ as multiplication by $z^{(a_i,r)}$. It acts as identity on the symmetric algebra. Taylor expand the exponentials to obtain

$$Q\left(e^{a_i}, z\right) a_j(1) = e^{a_i}\left(1 + a_i(1)z^1 + \ldots\right)\left(1 - a_i(-1)z^{-1} + \ldots\right) a_j(1)$$

The bracket $[e_i, h_j]$ is the coefficient of z^{-1} of the expansion. The terms $a_i(-n)a_j(1)$ vanish for $n \neq 1$. Hence, when applied to $a_j(1)$, only the term $e^{a_i} \times (1) \times (-a_i(-1)z^{-1})$ gives a non-zero contribution to the Lie-algebra bracket. Thus,

$$[e^{a_i}, a_j(1)] = -(a_i, a_j)e^{a_i}.$$

This provides an illustration of how the notions of the roots and root spaces of an ordinary Kac-Moody algebra are embedded in the framework of vertex algebras. If we define the natural L-grading of $V(L)$ as

(1.12) $$\deg_L(e^r) = r, \ \deg_L(t(p)) = 0,$$

then we find that the Lie algebra P^1/DP^0 has a natural decomposition into root spaces. Furthermore, for any root $r \in L$, the root space of r is the homogeneous subspace of the Lie algebra of degree r.

We can define a unique bilinear form on $V(L)$ through the following two requirements [**Bor86**]:

(1.13a) $$(e^{r_1}, e^{r_2}) = \begin{cases} 1 & \text{if } r_1 = -r_2, \\ 0 & \text{otherwise;} \end{cases}$$

(1.13b) $$\text{the adjoint of } t(p) \text{ is } -t(-p).$$

This bilinear form induces an invariant bilinear form on the Lie algebra P^1/DP^0 ([**Bor92**], proof of theorem 6.1). Evaluating the bilinear product on $V(L)$ we find that L_i is the adjoint of L_{-i}. Let $v \in P^0$, and $w \in P^1$. Then Dv is in P^1, and $(Dv, w) = (L_{-1}v, w) = (v, L_1 w) = 0$. The kernel of the bilinear form $(\cdot, \cdot)$ on P^1 thus contains DP^0.

Using this explicit construction of the Lie algebra elements, [**Bor86**] announces the following results about the Lie algebra P^1/DP^0.

THEOREM 1.4. *Let L be a non-singular, even lattice. Let the physical space $P^1 \subset V(L)$ be defined as in (1.7).*

a) Let A be a Lie algebra, A connected, simply laced, and not affine. If the lattice L contains the root lattice of A (possibly quotiented out by some null lattice) then A can be mapped to P^1/DP^0 such that the kernel is in the centre of A.

b) Let d be the dimension of L, and let $r \in L$ be a root such that $r^2 \leq 0$. Let $p_d(n)$ denote the number of partitions of n into d colours. Then the dimension of the degree r subspace of P^1/DP^0 is equal to

$$p_{d-1}\left(1 - \frac{r^2}{2}\right) - p_{d-1}\left(-\frac{r^2}{2}\right).$$

This forms an upper limit of the multiplicities of the roots of any Lie algebra A which satisfies the assumptions in a).

To provide an indication of the proof and the type of elements in the space P^1/DP^0, we recall from (1.10) that the degree r part of P^1 is spanned by vectors of the form $v = e^r \prod t_i(p_i)$. Being an element of P^1, v must be eigenvector of L_0 with eigenvalue 1, hence we require $1 = \frac{r^2}{2} + \sum p_i$. The sum over p_i defines a partition (with colours) of the total of $1 - \frac{r^2}{2}$. The colours of the partition correspond to the dimension d of the underlying lattice L - note that the $t_i(p_i)$ are not necessarily linearly independent. The conditions $L_i v = 0, i > 0$, of formula (1.7), introduce further restrictions which reduce the dimension d by 1. Thus, $p_{d-1}(1 - r^2/2)$ is the dimension of the space P^1. Similarly, $p_{d-1}(-r^2/2)$ is the dimension of P^0, and we note that D maps P^0 injectively into P^1. [**Bor86**] announces that the above formulas can be generalized for the case that A is not simply laced. We will not require the generalized formulas in this work, though. We are now in the position to quote Borcherds' version of the no-ghost theorem. (The original form of the no-ghost theorem was proven by Goddard and Thorn.)

THEOREM 1.5. *Suppose that V is a vector space with a non-singular bilinear form $(\cdot, \cdot)$ and suppose that V is acted on by the Virasoro algebra in such a way that the adjoint of L_i is L_{-i}, the central element of the Virasoro algebra acts as multiplication by 24, any vector of V is a sum of eigenvectors of L_0 with non-negative integral eigenvalues, and all the eigenspaces of L_0 are finite dimensional. We let V_n be the subspace of V on which L_0 has eigenvalue n. Assume that V is acted on by a group G which preserves all this structure. We let $V(II_{1,1})$ be the vertex algebra of the double cover of the two dimensional even unimodular Lorentzian lattice $II_{1,1}$ (so that $V(II_{1,1})$ is $II_{1,1}$-graded, has a bilinear form $(\cdot, \cdot)$ and is acted on by the Virasoro algebra as above). We let P^1 be the subspace of the vertex algebra $V \otimes V(II_{1,1})$ of vectors v with $L_0(v) = v$, $L_i(v) = 0$ for $i > 0$, and we let P_r^1 be the subspace of P^1 of degree $r \in II_{1,1}$. All these spaces inherit an action of G from the action of G on V and the trivial action of G on $V(II_{1,1})$ and $\mathbb{R}^2$. Then the quotient of P_r^1 by the nullspace of its bilinear form is naturally isomorphic, as a G module with an invariant bilinear form, to $V_{1-r^2/2}$ if $r \neq 0$ and to $V_1 \oplus \mathbb{R}^2$ if $r = 0$.*

In the next section we will describe how [**Bor90b**] used the no-ghost theorem to identify the root spaces of certain GKMs explicitly. Note that the theorem only applies in the case of central charge 24, which is one reason why Borcherds' construction cannot be generalized straightforwardly to lattices of dimension other than 24.

1.4. The Fake Monster Lie Algebra

We will now recall Borcherds' construction of the fake monster Lie algebra. The original calculations were published in [**Bor90b**], where the algebra is called the monster Lie algebra. The construction uses the Leech lattice, which is an even lattice of 24 dimensions, such that all the results of the previous section, including the no-ghost theorem, apply.

For this construction we consider the vertex algebras of three different lattices, each with their two respective natural gradings, as defined above in equations (1.11) and (1.12). The three lattices are

Λ - the Leech lattice,

$II_{1,1}$ - the unique even 2-dimensional unimodular Lorentzian lattice,

$II_{25,1}$ - the 26-dimensional unimodular Lorentzian lattice $\Lambda \oplus II_{1,1}$.

If L is any of the above three lattices and $V(L)$ the vertex algebra associated with L, we use the L-grading (1.12) and $\mathbb{Z}$-grading (1.11) to define $V(L)_r$, $V(L)_n$ and $V(L)_{(r,n)}$ as the parts of $V(L)$ of $\deg_L = r$ or $\deg_{\mathbb{Z}}^{(L)} = n$, respectively. We will also use S_n, where S is the symmetric algebra, introduced in formula (1.9), and the $\mathbb{Z}$-grading on S is the restriction of the $\mathbb{Z}$-grading on $V(L)$.

We define the norm of an element $(m,n) \in II_{1,1}$ as $-2mn$. Therefore, the norm of an element $(\lambda, m, n) \in II_{25,1}$ is $\lambda^2 - 2mn$. As introduced in equation (1.3), the norm is the square of the usual one.

We consider the physical space $P^1_{(V(\Lambda) \otimes V(II_{1,1}))}$ of $V(\Lambda) \otimes V(II_{1,1})$, and a general element (m,n) of $II_{1,1}$. Then $\left(P^1_{(V(\Lambda) \otimes V(II_{1,1}))}\right)_{(m,n)}$ is the part of $P^1_{(V(\Lambda) \otimes V(II_{1,1}))}$ whose $II_{1,1}$-grading is (m,n). On this space, a bilinear form is defined as in (1.13). Let K denote the null space of this bilinear form. The no-ghost theorem 1.5 identifies

$$\left(P^1_{(V(\Lambda) \otimes V(II_{1,1}))}\right)_{(m,n)}/K \stackrel{\sim}{=} V(\Lambda)_{1 - \frac{(m,n)^2}{2}} = V(\Lambda)_{1+mn}.$$

Using the Λ-grading of $V(\Lambda)$ on both sides, Borcherds refines the isomorphism to

$$\left(P^1_{(V(\Lambda) \otimes V(II_{1,1}))}\right)_{(\lambda,m,n)}/K \stackrel{\sim}{=} V(\Lambda)_{(\lambda,1+mn)}.$$

Next, consider the vertex algebra of the 26-dimensional Lorentzian lattice $V(II_{25,1})$ which is isomorphic to $V(\Lambda) \otimes V(II_{1,1})$ [**Bor92**]. Consider the physical space $P^1_{(V(II_{25,1}))}$. Let K' denote the kernel of the bilinear form $(\cdot, \cdot)$ on $P^1_{(V(II_{25,1}))}$. Then [**Bor90b**] defines the fake monster Lie algebra as

$$M_\Lambda = \left(P^1_{(V(II_{25,1}))}\right)/K' \stackrel{\sim}{=} \left(P^1_{(V(\Lambda) \otimes V(II_{1,1}))}\right)/K.$$

(The explicit construction first quotients out DP^0 but we know that $DP^0 \subset K'$.) Combining the two isomorphisms above, we obtain that

$$(M_\Lambda)_{(\lambda,m,n)} \stackrel{\sim}{=} V(\Lambda)_{(\lambda,1+mn)}.$$

Note that the left hand side is derived from the vertex algebra of the Lorentzian lattice $II_{25,1}$, and is graded in $II_{25,1}$. The right hand side is a piece of the vertex algebra of the Leech lattice with Λ and $\mathbb{Z}$ gradings.

[**Bor90b**] shows that the Lie algebra M_Λ satisfies all conditions of theorem 1.1 above. M_Λ is therefore associated with a GKM, through central extensions, and quotienting out the null space of the bilinear form. Section 5 of [**Bor90b**] identifies the universal central extension $\hat{M}$ of M_Λ. The no-ghost theorem provides us with an explicit description of the building blocks of M_Λ, in terms of subspaces of vertex algebras. [**Bor90b**] uses this, the denominator formula for GKMs (theorem 1.3), and the theory of modular forms to determine the simple roots of M_Λ.

The root lattice of the Lie algebra P^1/DP^0, as a subset of the dual of the (unspecialized) Cartan subalgebra, is infinite-dimensional and singular with regard to the scalar product. Quotienting out the null space K' corresponds to a specialization of the root space in the terminology of [**Jur96**] as in section 1.1.2, above. Theorem 1.3 provides a denominator formula for a suitably extended Lie algebra $\hat{M}^e$. Before we can formulate the denominator formula of M_Λ itself (that is, without any extension of the Cartan subalgebra) we need to verify the validity of the specialization. Let π_0 denote the projection from P^1/DP^0 to M_Λ, and equally the projection of the dual spaces of the Cartan subalgebras (that is, the spaces within which the roots are defined). The subscript 0 is used to indicate that the projection is defined through the nullspace of the scalar product. [**Bor92**] shows that, under the projection π_0, the set of roots is mapped into the lattice $II_{25,1}$. Defining the norm zero vector $\rho = (0,0,1) \in II_{25,1}$, [**Bor92**] establishes that the real simple roots are projected to $(\lambda, 1, \frac{\lambda^2}{2} - 1)$ with multiplicity 1 and the imaginary simple roots project to $n\rho = (0,0,n)$ with multiplicity 24 for all $n > 0$. We have $(\rho, \pi_0 r) = -\frac{r^2}{2}$ for all simple roots r, which makes ρ a Weyl vector. The set of positive roots projects to the set of vectors in $II_{25,1}$ of norm at most 2 which are either positive multiples of ρ or have negative inner product with ρ.

An explicit description of the images of all positive roots under the projection π_0 can be given as follows. A root r of M_Λ is positive if for $\pi_0 r = (\lambda, m, n)$ either $m > 0$, or $m = 0$ and $n > 0$ holds. A root is negative if either $m < 0$, or $m = 0$ and $n < 0$. Note that no element r with $\pi_0 r = (\lambda, 0, 0)$ can be a root as, in that case, $r^2 \geq 4$. Thus, the above description covers all roots of M_Λ.

Using the explicit description, it is straightforward to see that the projection π_0 does not create infinite terms, nor does it identify positive and negative roots. Evaluating the actual root multiplicities provided by Borcherds' denominator formula, we also find that the projection does not identify simple and non-simple roots.

Furthermore, the explicit identification of the roots then allows to build the matrix of scalar products which satisfies conditions (C1) to (C3) of a generalized Cartan matrix. The inner product with the Weyl vector can be used as a $\mathbb{Z}$-grading of M_Λ.

Having verified that the specialization of the roots to $II_{25,1}$ is both valid as an abstract identity, and meaningful as the denominator formula of M_Λ, we can apply the full theory of denominator formulas to M_Λ, with roots in $II_{25,1}$. We will write $M_{(\lambda,m,n)}$ for $(M_\Lambda)_{(\lambda,m,n)}$. Let furthermore E denote the part of M_Λ corresponding to the positive roots.

We can summarize the above results in the (specialized) denominator formula of M_Λ. The remark following theorem 1.3 identified $\epsilon(\cdot)$ in the case that the simple roots are not linearly independent, as occurs in this specialization. All positive multiples of ρ are simple roots of multiplicity 24 and are perpendicular to each other, so $|\epsilon(n\rho)|$ is $p_{24}(n)$, the number of partitions of n into parts of $24 = \dim(\Lambda)$ colours (notation introduced in theorem 1.4). Thus $\epsilon(n\rho)$ equals the coefficient of q^n in $\prod_n (1-q^n)^{24} = q^{-1}\eta^{24}(q)$ where η is the Dedekind eta-function. We will give details on η in chapter 2. [**Bor90b**] obtains

$$(1.14) \qquad e^\rho \prod_{r \in (II_{25,1})^+} (1 - e^r)^{p_{24}(1 - r^2/2)} = \sum_{w \in W} \det(w) w\big(\eta^{24}(e^\rho)\big).$$

Here, $(II_{25,1})^+$ is the set of positive roots identified above, and W is the Weyl group, which is the group of isometries of the root lattice generated by the reflections corresponding to the real simple roots. [**Bor90b**] gives a full description of the Cartan subalgebra of the universal central extension $\hat{M}$ of M_Λ, as introduced in section 1.1.4. An explicit basis for the Cartan subalgebra is the sum of a one dimensional space for each vector of the Leech lattice and a space of dimension $24^2 = 576$ for each positive integer n. The significance of the latter is that, for each n, the imaginary simple root $n\rho$ has multiplicity 24. Correspondingly, there are $24^2 = 576$ generators $h_{ij}, i,j = 1,\ldots 24$. Because of the nature of its construction, it would be tedious to describe the Cartan subalgebra of M_Λ itself. For details of the constructions, see [**Bor90b**].

We conclude this section with a brief remark about the monster Lie algebra. The name was in [**Bor90b**] used for the GKM which we now call fake monster Lie algebra. In [**Bor92**] the (proper) monster Lie algebra was constructed from the vertex algebra of the monster Lie group, as introduced in [**FLM88**]. The steps from the vertex algebra to the Lie algebra are parallel to the case of the fake monster Lie algebra, above. The vertex algebra is tensored with $V(II_{1,1})$, the physical spaces P_r^1 and the kernel K of the bilinear form are defined as above. Again, the no-ghost theorem 1.5 applies to the pieces P_r^1/K and provides an explicit description of the building blocks. [**Bor92**] uses this to show that the Lie algebra P^1/K satisfies all conditions of theorem 1.1, above, and hence is associated with a GKM. [**Bor92**] calculates an explicit denominator formula, analogue to formula (1.14). Again, this is a specialization in the terminology of [**Jur96**]. Theorem 6.1 of [**Jur98**] discusses the details of this specialization and identifies a GKM $\mathfrak{g}(M)$ and an ideal $\mathfrak{c}$ so that the Monster Lie algebra can be recovered as the quotient $\mathfrak{g}(M)/\mathfrak{c}$. In the specialization, the simple roots of the monster Lie algebra are identified as the set $\{(1,i) \mid i = -1, 1, 2, 3, \dots\}$. The steps to prove that the specialization is well defined are similar to the case of the fake monster Lie algebra because the set of projected simple roots of the monster Lie algebra displays characteristics very similar to the set of projected simple roots of the fake monster Lie algebra.

1.5. The Twisted Denominator Formula

[**Bor92**] uses the concept of 'twisted' denominator formulas, which are a generalization of the ordinary denominator formulas. Starting from the equality of graded virtual spaces (1.5), $\bigwedge(E) = H_*(E)$, we obtained the ordinary denominator formulas by calculating the formal character

$$\chi = \sum_\lambda (\dim V_\lambda) e^\lambda = \sum_\lambda (\text{Tr id}|\, V_\lambda) e^\lambda,$$

on both sides, as in (1.6). As before, V_λ corresponds to a lattice grading of V. Let σ be an automorphism on V. Then we may, more generally, calculate the trace of σ

(1.15) $$\sum_\lambda (\text{Tr }\sigma|\, V_\lambda) e^\lambda$$

on both sides of (1.5). We will arrive at some generalized ('twisted') denominator formula. In the case of the monster Lie algebra introduced in the concluding paragraph of section 1.4, we may choose σ to be an automorphism of the monster group. The resulting twisted denominator formula is closely related to the Thompson series $T_\sigma(q)$ (see [**Bor92**] for more details).

Returning to the fake monster Lie algebra, we consider automorphisms of the vertex algebra $V(\Lambda \oplus II_{1,1})$ which are induced by automorphisms of $\hat\Lambda$, that is the Leech lattice, centrally extended by a group of order 2, as defined in equation (1.8). If $\sigma \in \text{Aut}(\hat\Lambda)$ then we define, using the notation of equation (1.10) for a general element of $V(\Lambda \oplus II_{1,1})$,

(1.16) $$\sigma(v) := \sigma(e^r) \bar\sigma t_1(p_1) \dots \bar\sigma t_q(p_q).$$

1.5. THE TWISTED DENOMINATOR FORMULA

Here $\bar\sigma$ is the projected action on elements of Λ which is well defined because $\mathrm{Aut}(\hat\Lambda)$ is an extension of $\mathrm{Aut}(\Lambda)$ of the form $2^{24}\cdot\mathrm{Aut}(\Lambda)$, such that $\sigma(e^r) = \epsilon_\sigma(r)e^{\sigma r}$ where ϵ_σ satisfies equation (1.8). The '$\cdot$' indicates that the extension is nonsplit, cf. the Atlas [**Con85**]. Note in particular that $\deg_\Lambda(\sigma(\mathrm{v})) = \bar\sigma(\deg_\Lambda(\mathrm{v}))$. Now let σ be an automorphism of $\hat\Lambda$ of finite order N. Thus we can regard $\bar\sigma$ as an element of $SL_{24}(\mathbb{Z})$. Its Jordan normal form is diagonal and the eigenvalues $\epsilon_j, j = 1,\ldots, 24$ are n^{th} roots of unity where $n|N$. Equivalently, we can describe σ through its cycle shape $a_1^{b_1}\ldots a_s^{b_s}$. We associate σ to the following modular form:

$$(1.17) \qquad \eta_\sigma(q) := \eta(\epsilon_1 q)\ldots\eta(\epsilon_{24}q) = \eta(q^{a_1})^{b_1}\ldots\eta(q^{a_s})^{b_s}.$$

Here η stands for the Dedekind eta-function. We will give details on η in chapter 2.

Restricting further, let $\sigma \in \mathrm{Aut}(\hat\Lambda)$ be of prime order N, let Λ^σ denote the fixed point lattice and $L = \Lambda^\sigma \oplus II_{1,1}$ the corresponding Lorentzian lattice. Following [**Bor92**] (section 13, p. 438) we assume for simplicity that any power σ^n of σ fixes all elements of $\hat\Lambda$ which are in the inverse image (with respect to the projection $\hat\Lambda \to \Lambda$) of any vector of Λ fixed by σ^n. This final assumption will be satisfied by all automorphisms σ of interest in this work.

For the dual lattices we find $L^* = \Lambda^{\sigma*} \oplus II_{1,1}$. The $(II_{25,1})$-grading of M_Λ induces an L^*-grading as follows. It is well known that the projection $\pi_\sigma : \Lambda \to \Lambda^{\sigma*}$ maps onto the dual. (See theorem 3.1 below for a proof.) For $r = (\lambda^*, m, n) \in L^*$ let

$$(1.18) \qquad \tilde M_r = \tilde M_{(\lambda^*,m,n)} = \bigoplus_{\lambda \in \Lambda : \pi_\sigma(\lambda) = \lambda^*} M_{(\lambda,m,n)}.$$

Also, we will write $\tilde E_r$ and E_r for $\tilde M_r$ and M_r if we want to emphasize that we consider positive roots.

If we consider both sides of equation (1.5), $\bigwedge(E) = H_*(E)$, as L^*-graded $\mathrm{Aut}(\hat\Lambda)$ modules then we can calculate the trace of σ. For any $r \in L^*$ define the numbers $\mathrm{mult}(r)$ as

$$(1.19) \qquad \mathrm{mult}(r) = \sum_{d,s>0, ds|((r,L),N)} \frac{\mu(s)}{ds}\mathrm{Tr}(\sigma^d|\tilde E_{\frac{r}{ds}})$$

[**Bor92**], (13.2). Here $\mu(s)$ is the Möbius function, and (r, L) denotes the greatest common divisor of the numbers (r, a) for $a \in L$. Then the trace (1.15) on the equation (1.5) of graded virtual vector spaces is

$$(1.20) \qquad e^\rho \prod_{r \in L^{*+}} (1 - e^r)^{\mathrm{mult}(r)} = \sum_{w \in W^\sigma} \det(w) w(\eta_\sigma(e^\rho))$$

cf. [**Bor92**], (13.3), with η_σ as in (1.17) and $\rho = (0,0,1) \in L$. The proof is quite similar to the proof of theorem 1.3, that is the untwisted character formula. For the left hand side, again, the combinatorial argument applies. For the right hand side, again, the significant contribution is the term $w(\eta_\sigma(e^\rho))$, which is derived directly from the imaginary roots of M_Λ, as described in theorem 1.2b.

Equation (1.20) contains two terms, W^σ and L^{*+}, which need to be discussed in more detail. [**Bor92**] introduces the term W^σ to denote the subgroup of the Weyl group W of M_Λ consisting of reflections which commute with σ. This group is a subgroup of the reflection group of L. Theorem 2.2 of [**Bor90a**] provides a number of useful equivalent characterisations of the group W^σ. Let π_σ denote the projection from the span of $II_{25,1} = \Lambda \oplus II_{1,1}$ to the span of $L = \Lambda^\sigma \oplus II_{1,1}$ induced by the automorphism σ:

LEMMA 1.1. *Let $\sigma \in \mathrm{Aut}(\hat{\Lambda})$ be as above, let W be the Weyl group of M_Λ as introduced in equation (1.14). Let W^σ be the group of automorphisms introduced in equation (1.20). Then the following are equivalent characterizations of W^σ.*

(α) *W^σ is the group of elements of W commuting with σ.*
(β) *W^σ is the group of elements of W fixing the subspace L.*
(γ) *W^σ is generated by the reflections of the vectors $\pi_\sigma r$ as r runs through the simple roots of W whose projections $\pi_\sigma r$ have positive norm.*
(δ) *Same as (γ), with 'simple roots' replaced by 'roots'.*

PROOF. Theorem 2.2 of [**Bor90a**]. □

[**Bor92**] identifies L^{*+} as the set of all $r = (\lambda^*, m, n) \in L^*$ such that $m > 0$, or $m = 0$ and $n > 0$, which can be derived directly form the description of the positive roots of M_Λ in section 1.4, above. Hence, the projection π_σ does not identify positive and non-positive roots of M_Λ. [**Bor92**] observes that $\pi_\sigma \rho$ is a norm 0 Weyl vector for W^σ which we will again denote ρ for simplicity. The description (γ) of lemma 1.1, above, shows that the simple roots of W^σ can be represented through vectors in L^{*+}.

1.6. Construction of the GKMs

We continue to use the notation of the previous section. That is, let $\sigma \in \mathrm{Aut}(\hat{\Lambda})$ be of prime order N. Now assume furthermore that N is any one of 2, 3, 5, 7, 11, or 23, and that σ is of cycle shape $1^M N^M$ where $M = 24/(N+1)$. Lemma 12.1 of [**Bor92**] confirms that these automorphisms satisfy all assumptions placed on σ in the previous section 1.5. Let Λ^σ denote the fixed point lattice and $L = \Lambda^\sigma \oplus II_{1,1}$ the corresponding Lorentzian lattice as above. Starting from equation (1.20), the aim of this section will be to construct a new generalized Cartan matrix, and thus a new GKM for each N. [**Bor92**] identifies a subset of L^{*+} which constitutes the set of 'prospective simple roots'. Clearly, they are not linearly independent. Therefore, we expect that they will be roots resulting from a specialization in the sense of [**Jur98**], and that we will recover (1.20) as the denominator formula of the new (specialized) GKM.

Before we can verify the properties of a generalized Cartan matrix we need to recall a few results of [**Bor90a**] about the fixed point lattice L and the Weyl group W^σ. As seen in lemma 1.1, the group W^σ is a subgroup of the reflection group of the fixed point lattice L. In general, it will not be the full reflection group of L. One obvious necessary condition for it to be the full group is that the lattice Λ^σ has no roots. Any root $\lambda \in \Lambda^\sigma$ induces a root $(\lambda, 0, 0)$ of L. However, in the cases considered in this work, the condition is also sufficient. The reflection induced by a root $r \in L$ is the same as that induced by nr where n is any non-zero integer. Hence we may restrict ourselves to roots r which are primitive in the sense that, if $n \in \mathbb{Z}, |n| > 1$, then $\frac{r}{n}$ is not in L.

LEMMA 1.2. *Let σ be an automorphism of the Leech lattice Λ such that the sublattice Λ^σ fixed by σ has no roots, and let $L = \Lambda^\sigma \oplus II_{1,1}$. Let $\rho = (0,0,1) \in II_{25,1} = \Lambda \oplus II_{1,1}$ denote the Weyl vector of $II_{25,1}$.*

a) The group W^σ is the full reflection group of the lattice L.

b) The simple roots of W^σ are exactly the roots $r \in L$ such that (r, ρ) is negative and divides (r, v) for all vectors v of L.

c) The vector ρ is also norm 0 Weyl vector for the lattice L. That is, the primitive simple roots of W^σ are exactly the primitive roots r of W^σ satisfying $(r, \rho) = -r^2/2$.

PROOF. The introduction of [**Bor90a**] asserts that all roots considered are implicitly understood to be primitive in that paper. [**Bor92**] uses the results of [**Bor90a**] so that the implicit assumption also applies to [**Bor92**]. Given this implicit understanding, claim b) is part of theorem 3.3 of [**Bor90a**]. Claim a) is an auxiliary result from the proof of the same theorem. Claim c) is quoted from section 13 of [**Bor92**]. □

In order to apply lemma 1.2 we still need to verify the assumptions about Λ^σ. These depend on specific properties of the Leech lattice and the cycle shape of σ. We will establish these properties in chapter 4.1, below. At this point, we merely state

LEMMA 1.3. *Let σ be an automorphism of the Leech lattice Λ of cycle shape $1^M N^M$ where N is any of the primes 2, 3, 5, 7, 11, 23, and $M = 24/(N+1)$. Then Λ^σ has no roots.*

PROOF. The proof will be provided in section 4.1, below.

Let us now return to the twisted denominator formulas. If we compare equation (1.20) to the denominator formula of a GKM (theorem 1.3 of section 1.2 above) we observe that the right hand side of (1.20) is precisely the right hand side of (the specialization of) a denominator formula for a GKM with Weyl group W^σ. Until the definition of the GKM and the verification of the specialization have been completed, we will need to distinguish carefully between the linearly independent roots of the suitably extended Lie algebra, and those of the specialization to L^*. We will refer to the linearly independent roots as 'roots', and to the (specialized) roots in L^* as 'images of roots'.

Assuming that there exists a suitable GKM with well-defined specialization, the images of its simple roots under the specialization must be selected in L^{*+} as follows:

(G1) As images of the real simple roots for the GKM, select the simple roots of the reflection group W^σ. From lemmas 1.1 and 1.2 it follows that each such simple root of W^σ may be represented through a vector r in L^{*+} within the fundamental Weyl chamber of W^σ. The scaling factor of the primitive representative r is such that $(r, \rho) = -r^2/2$.

(G2) As images of the imaginary simple roots, select the positive multiples $n\rho$ of the Weyl vector ρ, with multiplicities equal to

$$\text{mult}(n\rho) = \sum_{a_k \text{ divides } n} b_k, \quad n \in \mathbb{Z}, \ n > 0, \tag{1.21}$$

if σ has generalized cycle shape $a_1^{b_1} a_2^{b_2} \ldots$

The value for the multiplicity of imaginary simple roots specified in equation (1.21) can be obtained by carefully counting occurrences, using the general description of the space S of theorem 1.2b in section 1.2 - note that all multiples of ρ are mutually orthogonal to each other. Thus, all 'prospective simple roots' selected under (G1) and (G2) have been represented in the lattice L^{*+}. Let $\{r_i\}$ be the set of all such 'prospective simple roots'. A scalar product for this set is naturally inherited from the lattice. We form the matrix of scalar products $C = \bigl((r_i, r_j)\bigr)$.

LEMMA 1.4. *The matrix C satisfies the defining conditions (C1) to (C3) of a generalized Cartan matrix.*

PROOF. (C1) is clearly satisfied as the scalar product is symmetric.

(C2) We begin by considering matrix elements c_{ij} where either r_i or r_j are candidates for imaginary simple roots selected under (G2). All candidates for imaginary simple roots are positive multiples of $\rho = (0, 0, 1)$. For any element $r = (\lambda, m, n)$ of L^{*+}, the scalar product satisfies $(r, \rho) = -m \leq 0$. Now consider a matrix element c_{ij} where both r_i and r_j are candidates for real simple roots selected under (G1). By selection, the candidates are all part of the same fundamental Weyl chamber of W^σ. Hence, $(r_i, r_j) \leq 0$ for all i, j.

(C3) Let r_i be a candidate selected under (G1), that is $r_i^2 > 0$, and let r_j be any candidate. By selection $(r_i, r_i) = -2(r_i, \rho)$. Having verified all assumptions about Λ^σ in lemma 1.3, we may apply lemma 1.2. Therefore, the simple roots of W^σ are all in L (considered as a subset of L^*), rather than the whole of L^*. In particular, this holds for r_j. Furthermore, lemma 1.2 shows that the simple roots of W^σ are exactly the roots $r \in L$ such that (r, ρ) is negative and divides (r, v) for all $v \in L$.

Hence, we have for $v = r_j$,

$$2\frac{c_{ij}}{c_{ii}} = 2\frac{(r_i, r_j)}{(r_i, r_i)} = -\frac{(r_i, r_j)}{(r_i, \rho)} \in \mathbb{Z}.$$

□

The explicit calculation of all Cartan matrix elements will be carried out in section 5.1 (theorem 5.3). This will be simplified by a deeper understanding of the fixed point lattices involved, which we will gain in chapter 4.1 (which also contains the proof of lemma 1.3).

For the present argument, however, the existence of the generalized Cartan matrix suffices. Starting from the Cartan matrix constructed above, we may now construct the GKM $G_N = G_N(C)$ with generators e_i, h_i, and f_i and relations (R1) to (R5) where N is the order of the automorphism σ. Quotient out the null space of the bilinear form, and denote the resulting Lie algebra $\mathcal{G}_N$. We may extend the GKMs G_N by the standard set of outer derivations, to obtain a new suite of Lie algebras $\mathcal{G}_N^e$. The results of [**Jur96**] apply and from equation (1.2) it follows that the Lie algebra $\mathcal{G}_N^e$ has the decomposition $\mathcal{G}_N^e = E_N \oplus H_N^e \oplus F_N$. Note that the outer derivations are part of the null space of the bilinear form.

Consider the linearly independent simple roots of $\mathcal{G}_N^e$ in the extended root space $H_N^{e\,*}$. From theorem 1.3, we obtain a denominator formula in the extended root space. Recall that we began the construction by selecting candidates for simple roots in L^*. In order to establish that a specialization to L^* is meaningful we need to check that all terms remain finite. Let π_0 again denote the projection $\pi_0 : \mathcal{G}_N^e = E_N \oplus H_N^e \oplus F_N \to \mathcal{G}_N = E_N \oplus H_N \oplus F_N$, defined by quotienting out the null space of the bilinear form. We will use the term π_0 also for the corresponding map of the dual spaces of the Cartan subalgebras, that is, $\pi_0 : H_N^{e\,*} \to H_N^*$. By construction, we have $H_N^* \tilde{=} [L^*]$ where $[L^*] = \mathbb{R} \otimes L^*$ denotes the span of L^*. Recall that, for the duration of the verification of the specialization, we refer to the roots of $\mathcal{G}_N^e$ in $H_N^{e\,*}$ as 'roots'. We denote the (specialized) roots of $\mathcal{G}_N$ in L^* as 'images of roots'. Let π_σ denote the projection $\pi_\sigma : \Lambda \oplus II_{1,1} \to L = \Lambda^\sigma \oplus II_{1,1}$ induced by σ. We observe that the $II_{1,1}$ components are not affected by the projection π_σ.

The selection rules (G1) and (G2) imply that only finitely many simple roots are identified by the projection π_0. For the image of a general root, recall that L^{*+} is the set of all $(\lambda, m, n) \in L^*$ such that $m > 0$, or $m = 0$ and $n > 0$. Consider a general positive root r of $\mathcal{G}_N^e$ and its decomposition into simple roots r_i

$$\pi_0 r = (\lambda, m, n) = \sum_{i=1}^{p} \pi_0 r_i = \sum_{i=1}^{p} (\lambda_i, m_i, n_i)$$

where the r_i are $p > 0$ not necessarily distinct simple roots such that $\pi_0 r_i = (\lambda_i, m_i, n_i) \in L^{*+}$. The explicit description of the set L^{*+} yields that $m_i \geq 0$ for the images of all simple roots under π_0. If $m_i = 0$ then we find $n_i > 0$ directly from the description of L^{*+}. If $m_i > 0$ then, for the images of all real simple roots, the relation

$$\frac{1}{2}(\lambda_i^2 - 2m_i n_i) = \frac{1}{2}(\pi_0 r_i)^2 = -(\pi_0 r_i, \rho) = m_i$$

implies $n_i = \lambda_i^2/2m_i - 1 \geq -1$. These constraints imply that r must be the sum of at most m simple roots such that $m_i > 0$ and at most $m+n$ simple roots such that $m_i = 0$. Conversely, the constraints $m_i \geq 0$ and $n_i \geq -1$ imply that each simple root r_i must satisfy the conditions that $m_i \leq m$, $n_i < m+n$ and $\lambda_i^2 \leq 2m(m+n+1)$. Hence, for the images of any roots of $\mathcal{G}_N^e$ in L^* we have shown;

(α) Only finitely many simple roots are identified by the projection π_0.
(β) For all positive roots r, the maximum number of summands p, within any decomposition, is finite.
(γ) For all positive roots r, there exists a finite set of simple roots $S_r = \{r_i\}$ so that no $r_j \notin S_r$ may be part of any decomposition of the image of the root r.

Hence, all multiplicities in the specialized denominator formula will remain finite, and the formula is well defined. Having confirmed that the specialization is valid we conclude that the specialization of the denominator formula of the Lie algebra $\mathcal{G}_N^e$ to L^* is the denominator formula of the Lie algebra $\mathcal{G}_N$. By construction, the specialized denominator formula is the twisted denominator formula (1.20) we started with. As before, we extend the term GKM to Lie algebras which differ from a GKM in the strict sense of defining relations (R1) to (R5) in outer derivations and central extensions and quotients only. For the remainder of this work, we will make the distinction implicit, and refer to the *specialized* Lie algebra $\mathcal{G}_N$ with roots in L^* as the GKM $\mathcal{G}_N$. We have thus proven the following theorem announced in [**Bor92**] (the outstanding proof of lemma 1.3 will be given in chapter 4.1).

THEOREM 1.6. *Suppose N is any of the primes 2, 3, 5, 7, 11, or 23, such that $(N+1)$ divides 24. Suppose that $\sigma \in \mathrm{Aut}(\hat{\Lambda})$ is of order N and cycle shape $1^M N^M$ where $M = \frac{24}{N+1}$. Let $\rho = (0, 0, 1)$ denote the Weyl vector of the reflection group W^σ. Then there exists a GKM $\mathcal{G}_N$ with simple roots as follows;*

(1) The real simple roots are the simple roots of the reflection group W^σ in L^*, which are the roots r with $(r, \rho) = -(r, r)/2$. That is, ρ is also Weyl vector for the GKM $\mathcal{G}_N$.

(2) *The imaginary simple roots are the positive multiples $n\rho$, with multiplicities equal to*

$$\mathrm{mult}(n\rho) = \sum_{a_k \text{ divides } n} b_k, \quad n \in \mathbb{Z}, \ n > 0,$$

if σ has generalized cycle shape $a_1^{b_1} a_2^{b_2} \ldots$.

The denominator formula of $\mathcal{G}_N$ is given by (1.20). □

This construction of Cartan matrices and GKMs may be performed for more general automorphisms than the suite of σ considered above. The left hand side of equation (1.20) thus gives the multiplicities of the positive roots of this algebra. As these multiplicities could possibly be negative the algebra will, in general, be a Lie superalgebra.

1.7. Root Multiplicities

We continue to use the notation of the previous section. That is, let N be any one of the primes 2, 3, 5, 7, 11, 23. Let $\sigma \in \mathrm{Aut}(\hat{\Lambda})$ be of prime order N and cycle shape $1^M N^M$ where $M = 24/(N+1)$. Let Λ^σ denote the fixed point lattice and $L = \Lambda^\sigma \oplus II_{1,1}$ the corresponding Lorentzian lattice as above. We will evaluate formula (1.19)

$$\mathrm{mult}(r) = \sum_{ds \mid ((r,L), N)} \frac{\mu(s)}{ds} \mathrm{Tr}(\sigma^d | \tilde{E}_{\frac{r}{ds}})$$

to find explicit formulas for the root multiplicities. We recall that in equation (1.19) the brackets $(\cdot, \cdot)$ denote the greatest common divisor. N divides (r, L) if and only if $r \in NL^*$, otherwise we conclude $((r, L), N) = 1$. Thus the pair (d, s) can only be one of the following: $(1, 1), (1, N), (N, 1)$. The values of the Möbius function are $\mu(1) = 1$ and $\mu(N) = -1$ because N is prime. Using that the automorphism σ satisfies $\sigma^N = \mathrm{id}$ we obtain from (1.19)

(1.22) $$\mathrm{mult}(r) = \mathrm{Tr}(\sigma | \tilde{E}_r) - \frac{1}{N} \mathrm{Tr}(\sigma | \tilde{E}_{r/N}) + \frac{1}{N} \mathrm{Tr}(\mathrm{id} | \tilde{E}_{r/N})$$

We will determine $\mathrm{Tr}(\sigma | \tilde{E}_r)$ and $\mathrm{Tr}(\mathrm{id} | \tilde{E}_r) = \dim(\tilde{E}_r)$ for $r \in L^*$ in lemmas 1.6 and 1.7, however we first require

LEMMA 1.5. *Let V, W be $\mathbb{Z}$-graded vector spaces, σ, τ automorphisms of V, W respectively. Let $V \otimes W$ be graded by the tensor grading and define the generating function*

$$\phi_{\sigma, V}(q) = \sum_n \mathrm{Tr}(\sigma | V_n) q^n$$

and analogously for τ, W and $\sigma \otimes \tau, V \otimes W$. Then $\phi_{\sigma, V}(q) \, \phi_{\tau, W}(q) = \phi_{\sigma \otimes \tau, V \otimes W}(q)$.

PROOF. We observe that $\deg(v \otimes w) = \deg(v) + \deg(w)$. The proof now is a straightforward exercise. □

LEMMA 1.6. *$Tr(\sigma | \tilde{E}_r)$ is equal to the coefficient of $q^{(1-r^2/2)}$ in $q\eta_\sigma(q)^{-1}$ if $r \in L$ and zero otherwise.*

PROOF. If $r \notin L$ then it follows from the defining equation (1.18) that σ does not fix the individual roots $(\lambda, m, n) \in L$ which project to $r = (\pi_\sigma \lambda, m, n)$, hence the trace must be zero. The same argument proves for any $r \in L$ that

1.7. ROOT MULTIPLICITIES

$\text{Tr}(\sigma|\tilde{E}_r) = \text{Tr}(\sigma|E_r)$. If $r = (\lambda, m, n)$ then $E_r = e^{\lambda} S_{1+mn-\frac{\lambda^2}{2}}$. (Recall the $\mathbb{Z}$-grading of S from formula (1.11) and the construction of M_Λ in section 1.4.) Let x_j denote the eigenvector of σ of eigenvalue ϵ_j (see equation (1.17)) and let (i) indicate the specific copy of Λ. Define $S^{(j)} = S(\bigoplus_{i>0} x_{j(i)})$. For the complexified symmetric algebras we obtain

$$S = S\left(\bigoplus_{i>0} \Lambda_{(i)}\right) = \bigotimes_{j=1}^{24} S\left(\bigoplus_{i>0} x_{j(i)}\right) = \bigotimes_{j=1}^{24} S^{(j)}.$$

A vector space basis for the factor $S^{(j)}$ in the tensor product is given by all elements of the form $\prod_k x_{j(m_k)}^{n_k}$. Such an element has $\mathbb{Z}$-degree $\sum_k n_k m_k$ and eigenvalue $\epsilon_j^{\sum n_k}$. Thus the generating function for $S^{(j)}$ is the product over the collection of copies $\Lambda_{(i)}$:

$$\phi_{\sigma, S^{(j)}} = \prod_{i>0} \left(\sum_{n \geq 0} \epsilon_j^n q^{in}\right) = \prod_{i>0} \sum_{n \geq 0} (\epsilon_j q^i)^n = \prod_{i>0} \frac{1}{1 - \epsilon_j q^i}.$$

Thus the generating function of the trace is

$$(1.23) \qquad \phi_{\sigma, S}(q) = \prod_j \phi_{\sigma, S^{(j)}}(q) = \prod_j \prod_{i>0} (1 - \epsilon_j q^i)^{-1} = q \eta_\sigma(q)^{-1}$$

As $1 - \frac{r^2}{2} = 1 + mn - \frac{\lambda^2}{2}$ we can read off the relevant trace as claimed. $\square$

LEMMA 1.7. *Let $r = (\lambda^*, m, n) \in L^*$ and let $\lambda^{\perp *} \in \Lambda^{\sigma \perp *}$ such that $\lambda^* + \lambda^{\perp *} \in \Lambda$. The dimension of $\tilde{E}_r$ is equal to the coefficient of $q^{1-r^2/2}$ in $\frac{q}{\Delta(q)} \theta_{\Lambda^{\sigma \perp} + \lambda^{\perp *}}(q)$ where $\theta_{\Lambda^{\sigma \perp} + \lambda^{\perp *}}(q)$ is the theta-function of the translated lattice $\Lambda^{\sigma \perp} + \lambda^{\perp *}$.*

REMARK. For a precise definition of θ-functions, see chapter 3 below. We will use the shorter notation $\theta_{\lambda^{\perp *}}$ in the sequel.

PROOF.

$$\tilde{E}_r = \bigoplus_{\pi_\sigma \lambda = \lambda^*} E_{(\lambda, m, n)} = \bigoplus_{\pi_\sigma \lambda = \lambda^*} V_{(\lambda, 1+mn)}$$

$$= \bigoplus_{\pi_\sigma \lambda = \lambda^*} \mathbb{R}(\hat{\Lambda})_\lambda \otimes S_{1+mn-\frac{\lambda^2}{2}}$$

Now $1 - \frac{r^2}{2} = 1 + mn - \frac{(\lambda^*)^2}{2}$. Hence we will find $\dim(\tilde{E}_r)$ as the coefficient of $q^{1-\frac{r^2}{2}}$ in the series

$$\sum_p \sum_{\pi_\sigma \lambda = \lambda^*} \dim\left(S_{p + \frac{(\lambda^*)^2}{2} - \frac{\lambda^2}{2}}\right) q^p = \sum_{\pi_\sigma \lambda = \lambda^*} \sum_{p'} \dim(S_{p'}) q^{p' + \frac{\lambda^2}{2} - \frac{(\lambda^*)^2}{2}} =$$

$$= \sum_{\pi_\sigma \lambda = \lambda^*} q^{\frac{\lambda^2}{2} - \frac{(\lambda^*)^2}{2}} \sum_{p'} \dim(S_{p'}) q^{p'} = \theta_{\lambda^{\perp *}} \frac{q}{\Delta(q)}.$$

Here, $\Delta(q) = \eta(q)^{24}$ and $\frac{q}{\Delta(q)}$ is well known to be the generating function for the symmetric algebra (it describes the number of partitions into 24 colours, see [**Bor92**], section 12). This completes the proof of lemma 1.7. $\square$

Using the modular form η_σ as defined in equation (1.17) we define

(1.24) $$\sum_{j>0} p_\sigma(1+j) q^{1+j} = q/\eta_\sigma(q).$$

This is a generalized partition function. Using the Dedekind η-function we define for all elements τ of the upper half plane $\mathcal{H}$ the function

$$\psi_j(\tau) = \eta(\frac{\tau+j}{N} + j).$$

Setting $q = e^{2\pi i\tau}$, we also define $\psi_j(q)$ for $|q| < 1$. (For more details see chapter 2.2, especially equation (2.5) below.) Let $\delta(r \in L)$ be defined to be of value 1 if $r \in L$ and of value 0 otherwise. We are now in the position to state the central theorem

THEOREM 1.7. *Suppose N is any of the primes 2, 3, 5, 7, 11, or 23, such that $(N+1)$ divides 24. Suppose that $\sigma \in Aut(\hat{\Lambda})$ is of order N and cycle shape $1^M N^M$ where $M = \frac{24}{N+1}$. Then*

(1.25)
$$\theta_{\lambda^{\perp *}}(q) = \eta(q)^{NM}\left(\eta(q^N)^{-M}\delta(\lambda^{\perp *} \in \Lambda^{\sigma\perp}) + \sum_{0\leq j<N} e^{-j(\lambda^{\perp *})^2\pi i}\psi_j(q)^{-M}\right)$$

and the twisted denominator formula (1.20) has the following explicit form:

(1.26)
$$e^\rho \prod_{r\in L^+}(1-e^r)^{p_\sigma(1-r^2/2)} \prod_{r\in NL^{*+}}(1-e^r)^{p_\sigma(1-r^2/2N)}$$
$$= \sum_{w\in W^\sigma} \det(w) w \left(e^\rho \prod_{i>0}(1-e^{i\rho})^M(1-e^{Ni\rho})^M\right)$$

PROOF. The proof of (1.25) will be given in chapter 4. Assuming (1.25), we proceed to prove the denominator formulas (1.26). The right hand side follows straightforwardly from equation (1.20) and the definition of η_σ. For the left hand side we compare the multiplicities given by (1.22) with those claimed in the theorem. We distinguish 4 cases

Case 1: $r \notin L$. Obviously the multiplicities are 0. (Lemma 1.6)

Case 2: $r \in L, r \notin NL^*$. In (1.22) only the first summand is nonzero, lemma 1.6 proves the claimed formula.

Case 3: $r \in NL^*, r \notin NL$. Suppose $r = (N\lambda^*, m, n)$. We choose $\lambda^{\perp *}$ as in lemma 1.7. Under the assumptions of case 3 the first and third term of (1.22) are nonzero. The first accounts for the exponent $p_\sigma(1-r^2/2)$. By lemma 1.7, dim $\tilde{E}_{r/N}$ is the coefficient of $q^{1-\frac{r^2}{2N^2}}$ in $\frac{q}{\Delta(q)}\theta_{\lambda^{\perp *}}(q)$ or equivalently the coefficient of $q^{1-\frac{r^2}{2N}}$ in $\frac{q}{\Delta(q^N)}\theta_{\lambda^{\perp *}}(q^N)$.

In case 3, $r/N \notin L$ and we obtain from (1.25)

(1.27) $$\frac{1}{N}\frac{q}{\Delta(q^N)}\theta_{\lambda^{\perp *}}(q^N) = \frac{1}{N}q\eta(q^N)^{-M}\sum_{0\leq j<N}e^{-j(r/N)^2\pi i}\eta(\epsilon^j q)^{-M}.$$

Now

$$\sum_{0\leq j<N} e^{-j(\frac{r}{N})^2\pi i}\epsilon^{j\frac{r^2}{2N}} = \sum_{0\leq j<N} e^{-j(\frac{r}{N})^2\pi i}\left(e^{\frac{2\pi i}{N}}\right)^{j\frac{r^2}{2N}} = N.$$

1.7. ROOT MULTIPLICITIES

Thus the coefficient of $q^{1-r^2/2N}$ in (1.27) is equal to that of $q\eta(q^N)^{-M}\eta(q)^{-M} = q\eta_\sigma^{-1}(q)$. This completes case 3.

Case 4: $r \in NL$. Here all three terms of (1.22) are nonzero. As $\frac{r}{N} \in L$, we have an additional term in $\frac{q}{\Delta}\theta$, namely $q\eta(q^N)^{-M}\eta(q^{N^2})^{-M} = q\eta_\sigma^{-1}(q^N)$. Thus it cancels exactly with the contribution of $\text{Tr}(\sigma|\tilde{E}_{r/N})$. Now the argument is the same as in case 3. □

COROLLARY. *Let N be any of 2, 3, 5, 7, 11, 23. Then the generalized Kac-Moody algebras $\mathcal{G}_N$ constructed in theorem 1.6 have root lattice $L = \Lambda^\sigma \oplus II_{1,1}$. Their root multiplicities are functions of the norm of the roots as follows:*

$$\text{mult}(r) = p_\sigma(1 - \frac{r^2}{2}) \qquad r \in L,\ r \notin NL^*$$

and

$$\text{mult}(r) = p_\sigma(1 - \frac{r^2}{2}) + p_\sigma(1 - \frac{r^2}{2N}) \qquad r \in NL^*.$$

PROOF. The explicit multiplicities for any root follow by comparison of theorem 1.7, equation (1.26) with the general denominator formula theorem 1.3. □

We conclude the chapter with a number of remarks:

1.) In chapter 5 below we will identify the real simple roots of the GKM $\mathcal{G}_N$, as the following:

$$\left(\lambda, 1, \frac{\lambda^2}{2} - 1\right) \quad \text{for} \quad \lambda \in \Lambda^\sigma$$

and

$$\left(\lambda, N, \frac{\lambda^2}{2N} - 1\right) \quad \text{for} \quad \lambda \in N\Lambda^{\sigma*} \text{ such that } N \mid \left(\frac{\lambda^2}{2N} - 1\right).$$

The former real simple roots have height 1 and norm 2 whereas the latter have height N and norm $2N$. The imaginary simple roots have already been identified explicitly in theorem 1.6.

2.) Equation (1.26) can be read as a new combinatorial identity, independent of the theory of generalized Kac-Moody algebras, in the same way that the denominator formulas of affine Kac-Moody algebras give the Macdonald identities.

3.) The explicit root multiplicities form upper bounds for the root multiplicities of any subalgebras of the $\mathcal{G}_N$. This will be used in chapter 6.

4.) Finally, we will briefly discuss the limitations of the main methods of constructing GKMs outlined in this chapter. This will emphasize the pivotal role of the no-ghost theorem in the argument. The method of construction may be summarised in the following diagram where V symbolizes the construction of vertex algebra and physical spaces, K symbolises the operation of quotienting out the kernel of the bilinear form, and σ the application of the projection defined by the automorphism of order N:

$$\Lambda \xmapsto{V} P^1/DP^0 \xmapsto{K} M_\Lambda \xmapsto{\sigma} \mathcal{G}_N.$$

For the fake monster Lie algebra, [**Bor92**] presents the following results
  a) the Cartan matrix (identifying generators and relations),
  b) all root multiplicities,
  c) an explicit construction (as subspace V_{1+mn} of a vertex algebra).

We have seen in section 1.3 above that the result c) considerably simplifies the identification of Lie algebra elements, allowing to calculate explicit root multiplicities (theorem 1.4). This work provides results analogous to a) (all elements of the generalized Cartan matrix will be calculated in theorem 5.3) and b) (corollary to theorem 1.7) for the new family of GKMs $\mathcal{G}_N$. We do not know a natural construction for these GKMs.

It is natural to try and apply our construction method to other lattices, in particular, we may consider the sequence of operations symbolised in the following diagram;

$$\Lambda \xmapsto{\sigma} \Lambda^\sigma \xmapsto{V} P^1(N)/DP^0(N) \xmapsto{K} M_{\Lambda^\sigma}(N).$$

Consider the Lie algebras $P^1(N)/DP^0(N)$ first. Theorem 1.4, above, applies and provides us with the root multiplicities of these Lie algebras. Turning to $M_{\Lambda^\sigma}(N)$, [**Bor92**] reports that the bilinear form can be shown to be almost positive definite (the kernel K is empty if the dimension of the lattice Λ^σ is less than 24), which makes this Lie algebra a GKM. However, the no-ghost theorem 1.5 does not apply. Therefore, in this case we have neither an explicit construction of the simple roots, nor can we express any multiplicity formulas for the simple roots.

Neither $\mathcal{G}_N \subset M_{\Lambda^\sigma}(N)$ nor $M_{\Lambda^\sigma}(N) \subset \mathcal{G}_N$ hold. The first relation can be disproven by considering that $\mathcal{G}_N$ contains real roots of norm $2N$. Hence, their $\mathbb{Z}$-grading is N, and they will not be elements of $P^1(N)$. Consequently, they will not be element of $M_{\Lambda^\sigma}(N)$ either.

For the second relation, we consider the case $N = 23$. The hyperbolic algebra AE_3, as defined in the introduction, can be embedded into both $\mathcal{G}_{23}$ and $M_{\Lambda^\sigma}(23)$. (For details see chapter 6.2.1, below.) We consider its unique root of norm -2 (up to Weyl automorphisms), which may be expressed in terms of simple roots as $r = (2, 2, 1)$. Its multiplicity in AE_3 is 2. The corollary to theorem 1.7 shows that the multiplicity of r in $\mathcal{G}_{23}$ is also 2. However, $\dim(P_r^1) = p_2(2) = 5$, and $\dim(P_r^0) = p_2(1) = 2$. Section 1.3 identifies the explicit vertex algebra bases for the space P_r^1, and its bilinear form. Using these, it is straightforward to calculate that the kernel of the bilinear form (1.13) on P_r^1 (paired with P_{-r}^1) is equal to the 2-dimensional DP_r^0. Hence the multiplicity of r in $M_{\Lambda^\sigma}(23)$ is 3.

The considerations of the previous paragraph do not only disprove any inclusions but also indicate that upper bounds for the root multiplicites of AE_3 provided by $M_{\Lambda^\sigma}(23)$ will be inferior to those of $\mathcal{G}_{23}$.

CHAPTER 2

Modular Forms

This chapter is the first in a series of three chapters which give the proof of the central theorem 1.7. In section 2.1 we recall the definitions of the modular group, its subgroups, and of modular forms. We proceed to recall some properties of modular forms. The material can be found in any standard treatment of modular forms and it has been included here solely to set up the notation. Section 2.2 defines a number of modular forms related to the η-function. Their modularity properties are fully understood in principle. We will, however, need to know their exact characters explicitly in chapter 4. Therefore, section 2.2 will establish the transformation behaviour under the generators S and T of the modular group.

2.1. Review of Modular Group and Modular Forms

As all the material included in this section is standard textbook material we will not always give specific references. Let $\Gamma = SL_2(\mathbb{Z})$ denote the modular group, $\Gamma(N)$ its subgroup consisting of all elements which are identical to the identity matrix modulo N and

$$\Gamma_0(N) = \{\begin{pmatrix} a & b \\ c & d \end{pmatrix} \in \Gamma \mid c \equiv 0 \pmod{N}\}.$$

Recall that $\Gamma(N)$ and $\Gamma_0(N)$ are subgroups of the modular group of finite index. Furthermore let

$$S = \begin{pmatrix} 0 & -1 \\ 1 & 0 \end{pmatrix}, \qquad T = \begin{pmatrix} 1 & 1 \\ 0 & 1 \end{pmatrix}, \qquad F = \begin{pmatrix} 0 & -1 \\ N & 0 \end{pmatrix}.$$

F is called the Fricke involution. Then Γ is generated by T and S, $\Gamma_0(2)$ is generated by T and ST^2S, and for $N \neq 2$ it is known that $\Gamma_0(N)$ is generated by T and the collection of

(2.1) $$V_k = ST^k ST^{k'} S = \begin{pmatrix} -k' & 1 \\ kk' - 1 & -k \end{pmatrix}.$$

Here $k = 1, ..., N-1$ and k' is any integer such that $kk' \equiv 1 \pmod{N}$ (cf. [**Apo76**], theorem 4.3 or [**Rad29**]). $F^{-1} = \frac{-1}{N} F$ and

$$F \begin{pmatrix} a & b \\ Nc & d \end{pmatrix} F^{-1} = \begin{pmatrix} d & -c \\ -Nb & a \end{pmatrix}$$

In particular, if $kk' = N + 1$ then $FV_k F^{-1} = -V_{-k'}$. Let $\mathcal{H}$ denote the upper half complex plane, that is the set of all $\tau \in \mathbb{C}$ with positive imaginary part. Γ acts on $\mathcal{H}$ by $\begin{pmatrix} a & b \\ c & d \end{pmatrix}(\tau) = \frac{a\tau+b}{c\tau+d}$. The Fricke involution acts in the same way. Note that $SF(\tau) = N\tau$ and $FS(\tau) = \frac{\tau}{N}$.

The metaplectic group Mp is a double cover of Γ consisting of elements (A, j) where $A = \begin{pmatrix} a & b \\ c & d \end{pmatrix} \in \Gamma$ and j is a holomorphic function in $\tau \in \mathcal{H}$ such that

$j(\tau)^2 = c\tau + d$. The group multiplication is defined to be $(A, j(\tau))(B, k(\tau)) = (AB, j(B\tau)k(\tau))$. For any $k \in \frac{1}{2}\mathbb{Z}$ the metaplectic group then acts on the space of functions on $\mathcal{H}$ by

$$f|_{(A,j,k)}(\tau) = j(\tau)^{-2k} f(A(\tau)). \tag{2.2}$$

In particular, we will write $(\tau)_+^{1/2}$ to denote the branch of the square root that is positive for positive τ. $(S, +)$ will be the corresponding element of the metaplectic group using this branch. $(T, +)$ uses the constant square root of $+1$. Note that $(S, +)^2 = (-1, i)$. We will sometimes write $(G, *)$ when it is not required to specify the branch. Note further that $f^{-1}|_{(A,j,k)} = \left(f|_{(A,j,-k)}\right)^{-1}$. Let Γ' be a subgroup of Γ of finite index. An orbit in $\mathbb{Q} \cup i\infty$ under Γ' is called a cusp. Let Γ' be a subgroup of finite index of Γ of level N, that is, $\Gamma' \supset \Gamma(N)$. A function $f : \mathcal{H} \to \mathbb{C}$ is called a modular form of weight k and multiplier system χ for Γ' if f is holomorphic on $\mathcal{H}$ and $f|_{(A,j,k)} = \chi(A, j)f$ for all $A = \begin{pmatrix} a & b \\ c & d \end{pmatrix} \in \Gamma'$ and $\tau \in \mathcal{H}$. We can substitute $q = e^{2\pi i \tau}$, such that $|q| < 1$. f is called holomorphic modular form if in addition at each cusp of Γ' f has a q-expansion without negative powers.

We can now define the Dedekind eta-function. For $\tau \in \mathcal{H}$, let

$$\eta(\tau) = e^{\frac{2\pi i \tau}{24}} \prod_{1}^{\infty} (1 - e^{2\pi i n \tau}). \tag{2.3}$$

In terms of q we obtain the simpler form

$$\eta(q) = q^{\frac{1}{24}} \prod_{1}^{\infty} (1 - q^n). \tag{2.3'}$$

Note that (2.3') is well defined only by the additional prescription $q^{\frac{1}{24}} = e^{2\pi i \tau / 24}$. This will be implicitly understood whenever either variable q or τ is used in this work. We will also consider $\Delta(q) = \eta^{24}(q)$. It is well known that $\eta(S\tau) = e^{-\pi i/4}(\tau)_+^{1/2}\eta(\tau)$ and $\eta(T\tau) = e^{\pi i/12}\eta(\tau)$. Thus η is a modular form of weight $\frac{1}{2}$ and Δ is a holomorphic modular form of Γ of weight 12 and trivial multiplier system. Furthermore, Δ has a zero of first order at $q = 0$ and has no other zeros for $|q| < 1$.

Modular forms are uniquely determined by a certain number of initial coefficients. We will in particular use the following theorem.

THEOREM 2.1. *Let Γ' be a congruence subgroup of the modular group Γ of finite index. Let f be a holomorphic modular form of Γ' of integer weight k and a character χ such that $\chi^{12}(g) = 1$ for all $g \in \Gamma'$. Suppose that the q-expansions of $f(\tau)$ at all cusps do not contain any terms of q of exponent $\leq \frac{k}{12}$. Then $f \equiv 0$.*

PROOF. f^{12} is holomorphic in $\mathcal{H}$ and has trivial multiplier. By assumption its expansion at any cusp does not contain terms of exponent less or equal k. The q-expansion of Δ^k at any cusp has a kth order zero, hence $\frac{f^{12}}{\Delta^k}$ is a holomorphic modular form with trivial multiplier of Γ'. Its weight is 0. However, the only such forms are constant, cf. [**Kob84**], III.3.Prop.18. As the constant term in the q-expansion was assumed to be zero we conclude $f \equiv 0$. □

REMARK. The proof shows that a similar theorem holds for forms of any character of finite order.

COROLLARY. *Let N be a prime and $\Gamma' = \Gamma_0(N)$. Let f be as above. If the q-expansions of both f and $f|_S$ at zero do not contain any terms of exponent $\leq \frac{k}{12}$ then $f \equiv 0$.*

PROOF. $\Gamma_0(N)$ has exactly two cusps, represented by 0 and ∞, cf [**Shi71**], p.26. As S interchanges these two cusps the given criterion is equivalent to that of the theorem. □

2.2. Some Modular Forms Related to Eta

For the remainder of this chapter let N be an integer such that $N+1$ divides 24, and let $M = \frac{24}{N+1}$. We will need to know the transformation properties under the modular group for a number of functions related to the η-function. We begin by considering $\eta(N\tau)$, which corresponds to $\eta(q^N)$ in terms of the argument q. Then

(2.4a)
$$\eta(N\tau)|_{(S,+,\frac{1}{2})} = \left((\tau)_+^{1/2}\right)^{-1} \eta(NS\tau) = \left((\tau)_+^{1/2}\right)^{-1} \eta(S(\frac{\tau}{N})) = \frac{e^{-\pi i/4}}{\sqrt{N}} \eta(\frac{\tau}{N})$$

(2.4b) $\quad \eta(N\tau)|_{(T,+,\frac{1}{2})} = \eta(NT\tau) = \eta(T^N(N\tau)) = e^{\frac{N\pi i}{12}} \eta(N\tau)$

Now let

(2.5) $$\psi_j(\tau) = \eta(\frac{\tau+j}{N} + j)$$

which corresponds to $\psi_j(q) = \eta(\epsilon^j q^{1/N})$ in somewhat ambiguous notation where ϵ is an Nth root of unity. Note that these are the ψ_j as defined in chapter 1, and used in theorem 1.7.

(2.6) $\quad \psi_j|_{(T,+,\frac{1}{2})}(\tau) = \psi_j(T\tau) = \eta(\frac{\tau+(1+j)}{N} + (1+j) - 1) = e^{-\pi i/12} \psi_{j+1}(\tau).$

To determine $\psi_j^M|_{(S,+,\frac{1}{2})}$ we first notice that $\frac{\tau+(j+N)}{N} + (j+N) = \left(\frac{\tau+j}{N} + j\right) + (N+1)$. Now $\left(e^{\pi i/12}\right)^{(N+1)M} = 1$ and thus $\psi_{j+N}^M(\tau) = \psi_j^M(\tau)$. Given j, choose j' such that $jj' \equiv 1 (\text{mod} N)$ and define

(2.7) $\quad G = -FV_j F^{-1} = -FST^j ST^{j'} SF^{-1} = \begin{pmatrix} j & \frac{jj'-1}{N} \\ N & j' \end{pmatrix} \in \Gamma_0(N).$

The action of G on τ equals that of $-G$, hence we obtain that
$$\frac{S\tau + j}{N} + j = T^j FST^j S\tau = T^j GT^{j'} T^{-j'} FST^{-j'}\tau = T^j GT^{j'}\left(\frac{\tau - j'}{N} - j'\right)$$

Note furthermore that the holomorphic function j associated to the element G of equation (2.7) satisfies $j^2(\frac{\tau-j'}{N}) = \tau$. Hence

(2.8)
$$\psi_j|_{(S,+,\frac{1}{2})}(\tau) = \left((\tau)_+^{1/2}\right)^{-1} \psi_j(S\tau) = \left((\tau)_+^{1/2}\right)^{-1} \eta\left(T^j G(\frac{\tau-j'}{N})\right) =$$
$$= \left((\tau)_+^{1/2}\right)^{-1} e^{j\pi i/12} \eta\left(G(\frac{\tau-j'}{N})\right) = e^{j\pi i/12} \chi(G) \eta(\frac{\tau-j'}{N}) =$$
$$= e^{(j+j')\pi i/12} \chi(G) \psi_{-j'}(\tau).$$

The character $\chi(G)$ will have to be calculated case by case using the following two lemmata.

LEMMA 2.1. *Suppose that, for the element G of equation (2.7), $G = ST^a ST^b ST^c S$. Then*

$$\psi_j|_{(S,+,\frac{1}{2})}(\tau) = e^{-\pi i} e^{(j+j'+a+b+c)\pi i/12} i^J \psi_{-j'}(\tau)$$

Here $J = J(a,b,c)$ is an integer that will be specified in the proof.

PROOF. Let $y = \frac{\tau - j'}{N}$. Following the line of argument in formula (2.8) we find

$$\psi_j|_{(S,+,\frac{1}{2})}(\tau) = \left((\tau)_+^{1/2}\right)^{-1} e^{j\pi i/12} \eta(ST^a ST^b ST^c S(y)) =$$

$$= \left((\tau)_+^{1/2}\right)^{-1} e^{j\pi i/12} e^{-4\pi i/4} e^{(a+b+c)\pi i/12} * (y)_+^{1/2} \left(\frac{-1}{y} + c\right)_+^{1/2}$$

$$\left(\frac{-1}{\frac{-1}{y}+c} + b\right)_+^{1/2} \left(\frac{-1}{\frac{-1}{\frac{-1}{y}+c}+b} + a\right)_+^{1/2} e^{j'\pi i/12} \psi_{-j'}(\tau)$$

The factors containing τ or y other than η itself multiply to a constant as already seen in equation (2.8) above. To determine the correct branch of the square root it is sufficient to evaluate for $\tau = R + \frac{i}{R}$ where R is a large positive real. Then $\frac{1}{y}$ becomes arbitrarily small and it suffices to evaluate

$$(c)_+^{1/2} \left(\frac{-1}{c} + b\right)_+^{1/2} \left(\frac{-1}{\frac{-1}{c}+b} + a\right)_+^{1/2} = (c)_+^{1/2} \left(\frac{bc-1}{c}\right)_+^{1/2} \left(\frac{abc-a-c}{bc-1}\right)_+^{1/2}.$$

Note that a,b,c are integers and in particular real. If J is the number of changes of sign in the sequence $1, c, bc-1, abc-a-c = N$ then the correct branch is i^J. This completes and proves the claim. □

LEMMA 2.2. *Suppose that, in the above notation, $jj' = N+1$. Then $G = ST^{-j'}ST^{-j}S$ and*

$$\psi_j|_{(S,+,\frac{1}{2})}(\tau) = \operatorname{sign}(-j') e^{-3\pi i/4} \psi_{-j'}(\tau).$$

PROOF. The argument is a simplification of the proof of lemma 2.1. Here we have to count the number of sign changes in the sequence $1, -j', jj'-1 = N$. □

With these tools we can now describe the transformation of ψ_j^M for any j and N, M as above.

THEOREM 2.2. *Let N be any of 2, 3, 5, 11. Then*

$$\psi_j^M|_{(S,+,\frac{M}{2})}(\tau) = e^{\frac{-3\pi i}{4}M} \psi_{-j'}^M(\tau).$$

PROOF. Claim: We can use lemma 2.2 for all j and N. The cases $j = 1$ ($G = -ST^{-N}S$) and $j = -1$ ($G = ST^N S$) are trivial. We recall that we only need to consider j modulo N. Furthermore, as $jj' = (-j)(-j')$ it suffices to consider $j = 2, ..., \frac{N-1}{2}$. However, if $N = 5$, $2*3 = 5+1$, and if $N = 11$ we have the decompositions $jj' = 2*6 = 3*4 = (-6)*(-2) = 11+1$ (note that $-6 \equiv 5$ mod 11). Application of lemma 2.2 now gives the result as the factor $\operatorname{sign}(-j')$ is cancelled by the even power M. □

THEOREM 2.3. *Let $N = 7$ or $N = 23$. Then*

$$\psi_j^M|_{(S,+,\frac{M}{2})}(\tau) = e^{\frac{-3\pi i}{4}M} \left(\frac{-j'}{N}\right) \psi_{-j'}^M(\tau).$$

Here $\left(\frac{x}{y}\right)$ denotes the Legendre symbol.

2.2. SOME MODULAR FORMS RELATED TO ETA

PROOF. If $N = 7$ we use the decompositions $jj' = 2*4 = (-4)*(-2) = 7+1$ where $3 \equiv -4 \mod 7$. Squares modulo 7 are 1,2,4. Hence the claim follows from Lemma 2.2.

Now let $N = 23$. The squares modulo 23 are $1, 2, 3, 4, 6, 8, 9, 12, 13, 16, 18$. The decompositions $jj' = 1*24 = 2*12 = 3*8 = 4*6 = 23+1$ show which j are covered by Lemma 2.2. For the cases of the remaining j first observe that if

$$\psi_j^M|_{(S,+,\frac{M}{2})} = e^{\frac{-3\pi i}{4}M}\left(\frac{-j'}{N}\right)\psi_{-j'}^M$$

then

$$\psi_{-j'}^M|_{(S,+,\frac{M}{2})} = e^{\frac{3\pi i}{4}M}\left(\frac{-j'}{N}\right)\psi_{j}^M|_{(-1,i,\frac{M}{2})} = e^{\frac{3\pi i}{4}M}(-1)^M\left(\frac{j}{N}\right)i^{-M}\psi_j^M =$$
$$= e^{\frac{-3\pi i}{4}M}\left(\frac{j}{N}\right)\psi_j^M.$$

(Note that $\left(\frac{j}{N}\right) = \left(\frac{j'}{N}\right)$ as the Legendre symbol is multiplicative.) Thus we are only left to consider separately $j = 5, -5, 7, -7$ as we obtain $j' = -9, 9, 10, -10$.

Now $G = -FST^5ST^{-9}SF^{-1} = ST^{-5}ST^{-2}ST^2S$ hence by Lemma 2.1 we obtain

$$\psi_5|_{(S,+,\frac{1}{2})}(\tau) = e^{-\pi i}e^{(5+(-9)-5-2+2)\pi i/12}i^{J(-5,-2,2)}\psi_9(\tau)$$
$$= e^{-\pi i}e^{-9\pi i/12}(-1)\psi_9(\tau) = e^{-3\pi i/4}\psi_9(\tau)$$

As 9 is a square modulo 23 this is the claimed transformation.

The other cases are analogous, and it suffices to give the factor

$$e^{-\pi i}e^{(j+j'+a+b+c)\pi i/12}i^{J(a,b,c)}.$$

For $(j, j') = (-5, 9)$ we obtain $(a, b, c) = (4, -2, -3)$ and thus a factor of

$$e^{-\pi i}e^{(-5+9+4-2-3)\pi i/12}(-1) = e^{1\pi i/4} = e^{-3\pi i/4}\left(\frac{-9}{23}\right).$$

For $(j, j') = (7, 10)$ we obtain $(a, b, c) = (-3, 3, -2)$ and thus a factor of

$$e^{-\pi i}e^{(7+10-3+3-2)\pi i/12}(-1) = e^{5\pi i/4} = e^{-3\pi i/4}\left(\frac{-10}{23}\right).$$

For $(j, j') = (-7, 13)$ we obtain $(a, b, c) = (3, -4, -2)$ and thus a factor of

$$e^{-\pi i}e^{(-7+13+3-4-2)\pi i/12}(-1) = e^{1\pi i/4} = e^{-3\pi i/4}\left(\frac{-13}{23}\right).$$

This concludes the proof of theorem 2.3. $\square$

CHAPTER 3

Lattices and their Theta-Functions

This chapter establishes a number of properties of integral lattices which will be used in chapter 4. The lattices in question are those which are part of formula (1.25). Section 3.1 quotes a number of results concerning the dual of a sublattice of a self-dual lattice, and modularity properties of θ-functions under the generators S and T of the modular group. We then restrict our attention to the particular type of lattice that arises in (1.25). Again, the modularity properties of their θ-functions are fully understood in principle. Section 3.2 establishes the explicit modularity properties (including characters) of these θ-functions. Please note that in the notation of chapter 3, the letter L denotes a more general object than in chapter 1.

3.1. Review of Results about Lattices

For any lattice L, let $[L] = L \otimes \mathbb{R}$ denote the vector space spanned by the elements of L. Let Λ again denote the 24-dimensional unimodular self-dual Leech lattice. For any sublattice L of Λ, we define the orthogonal projection $\pi_L : [\Lambda] \to [L]$. We begin our review with a well known description for the dual of some lattices.

THEOREM 3.1. *Let L be a sublattice of Λ such that $\Lambda \cap [L] = L$. Define the projection $\pi_L : [\Lambda] \to [L]$ as above. Then $L^* = \pi_L \Lambda$.*

PROOF. The projection $\pi_L(\Lambda)$ is contained in the dual L^* because, for any $\lambda \in \Lambda$ and any $\mu \in L$, the inner products satisfy $(\pi_L \lambda, \mu) = (\lambda, \mu)$. The other inclusion will be proved by induction.

For any proper sublattice L_r of Λ such that $L_r = \Lambda \cap [L_r]$ we can choose $\lambda_{r+1} \in \Lambda$ such that $L_{r+1} := L_r \oplus \mathbb{Z}\lambda_{r+1}$ as lattices and that $L_{r+1} = \Lambda \cap [L_{r+1}]$. Below, we will show how, given $\lambda_r^* \in L_r^*$, to find $\lambda_{r+1}^* \in L_{r+1}^*$ such that $\pi_{L_r}(\lambda_{r+1}^*) = \lambda_r^*$. The proof can then be completed as follows: Given $\lambda^* \in L^*$ we choose $r_0 = \dim L$, $L_{r_0} = L$, $\lambda_{r_0}^* = \lambda^*$ and continue inductively until $L_{r_1} = \Lambda$. Then $\lambda_{r_1}^* \in \Lambda^* = \Lambda$ and hence $\lambda^* = \pi_L(\lambda_{r_1}^*) \in \pi_L(\Lambda)$ which finishes the proof.

We now construct λ_{r+1}^* for given $\lambda_r^* \in L_r^*$. Suppose L_r as a lattice has a basis $\lambda_1, ..., \lambda_r$. We choose rational $a_i, i = 1, ..., r$ such that $\lambda_{r+1} - \sum_{i=1}^{r} a_i \lambda_i \in [L_r^\perp]$. (This is possible because L is an integer lattice.) We use the ansatz

$$\lambda_{r+1}^* = \lambda_r^* + a\left(\lambda_{r+1} - \sum_{i=1}^{r} a_i \lambda_i\right)$$

with $a \in \mathbb{Q}$. We then have to satisfy the condition that the inner product of this expression with any element of L_{r+1} is integer. However, the inner product with any element of L_r is trivially integer. Thus we complete the proof by choosing a such that the inner product with λ_{r+1} is integer. □

We also quote the following fact from [**CS88**] (chapter 4, theorem 1):

THEOREM 3.2. *Let L be a sublattice of Λ. Let $L^\perp$ denote the orthogonal complement of L within Λ. Then the abelian group L^*/L is isomorphic to the group $(L^\perp)^*/(L^\perp)$.*

REMARK. Obviously theorems 3.1 and 3.2 remain true if we replace Λ by any self-dual lattice.

Let $\theta_{L+\lambda^*}$ denote the theta-function of the lattice L translated by λ^*, that is

$$(3.1) \qquad \theta_{L+\lambda^*}(q) = \sum_{\lambda \in L+\lambda^*} q^{(\lambda,\lambda)/2}.$$

As before, non-integer powers of $q = e^{2\pi i \tau}$ will be understood to be defined by $q^x = e^{2\pi i \tau x}$. We can now quote the following theorem which goes back to Jacobi and is a special case of theorem 13.5 of [**Kac90**]:

THEOREM 3.3. *Let L be an l-dimensional integral lattice, let λ^* be an element of L^*. Then*

$$\theta_{L+\lambda^*}|_{(S,+,\frac{l}{2})}(q) = \frac{e^{-l\pi i/4}}{\sqrt{|L^*/L|}} \sum_{\mu^* \in L^*/L} e^{-2\pi i(\lambda^*,\mu^*)} \theta_{L+\mu^*}(q)$$

and

$$\theta_{L+\lambda^*}|_{(T,1,\frac{l}{2})}(q) = e^{\pi i(\lambda^*,\lambda^*)} \theta_{L+\lambda^*}(q).$$

Furthermore, the action of the modular group on the vector space spanned by the functions $\theta_{L+\lambda^*}, \lambda^* \in L^*/L$ is unitary.

COROLLARY. *For $\lambda^* = 0$,*

$$\theta_L|_{(S,+,\frac{l}{2})}(q) = \frac{e^{-l\pi i/4}}{\sqrt{|L^*/L|}} \theta_{L^*}(q).$$

3.2. The Character of Theta

Let L be an even l-dimensional integer lattice with dual L^*. Let N be the least positive integer such that $NL^* \subseteq L$. Suppose that N is prime but not equal to 2. We calculate the transformation of θ_L under the generators of $\Gamma_0(N)$. Clearly θ_L is invariant under $(T,1)$. For $(V_k,*) = (ST^k ST^{k'} S, *)$ (cf equation (2.1)) we obtain

$$\theta_L\Big|_{\left((S,+)(T^k,1)(S,+)(T^{k'},1)(S,+),\frac{l}{2}\right)}$$

$$= \frac{e^{-l\pi i/4}}{\sqrt{|L^*/L|}} \sum_{\lambda \in L^*/L} \theta_{L+\lambda}\Big|_{\left((T^k,1)(S,+)(T^{k'},1)(S,+),\frac{l}{2}\right)} =$$

$$= \cdots = \frac{e^{-3l\pi i/4}}{\sqrt{|L^*/L|}^3} \sum_{\lambda,\mu,\nu \in L^*/L} e^{\pi i k(\lambda,\lambda) - 2\pi i(\lambda,\mu) + \pi i k'(\mu,\mu) - 2\pi i(\mu,\nu)} \theta_{L+\nu}$$

$$=: \sum_{\nu \in L^*/L} c_\nu(k) \theta_{L+\nu},$$

where this defines coefficients $c_\nu(k)$. We first evaluate the coefficient $c_0(k)$. Note that because $kk' \equiv 1 \pmod N$

$$k(\lambda - k'\mu)^2 = k\lambda^2 - 2kk'(\lambda,\mu) + k(k')^2\mu^2 \equiv k\lambda^2 - 2(\lambda,\mu) + k'\mu^2 \pmod{2\mathbb{Z}}.$$

Hence

$$c_0(k) = \frac{e^{-3l\pi i/4}}{\sqrt{|L^*/L|}^3} \sum_{\lambda,\mu \in L^*/L} e^{\pi i k(\lambda,\lambda) - 2\pi i(\lambda,\mu) + \pi i k'(\mu,\mu)} =$$

$$= \frac{e^{-3l\pi i/4}}{\sqrt{|L^*/L|}^3} \sum_{\lambda,\mu \in L^*/L} e^{\pi i k(\lambda - k'\mu)^2}$$

Now the sum runs through all elements of L^*/L exactly $|L^*/L|$ times. Thus

(3.2) $$c_0(k) = \frac{e^{-3l\pi i/4}}{\sqrt{|L^*/L|}} \sum_{\lambda \in L^*/L} e^{\pi i k \lambda^2}$$

L^*/L is an abelian group of, say, n generators x_i of order r_i, $i = 1, \ldots n$. There cannot be other relations. As we assume N is prime all r_i must equal N. Furthermore we note that $|L^*/L| = N^n$.

Hence any $\lambda \in L^*/L$ is of the form $\sum_{i=1}^{n} a_i x_i$ with $0 \le a_i \le N - 1$. Consider

$$\sum_{a_n=0}^{N-1} e^{\pi i k (y + a_n x_n)^2}$$

where y is a fixed element of L^*/L of the form $\sum_{i=1}^{n-1} a_i x_i$. Choose integers α, β such that $x_n^2 = \frac{2\alpha}{N}$, and $(x_n, y) \equiv \frac{2\beta}{N}(\mathbb{Z})$. Note that $\alpha \ne 0$. Choose α' such that $\alpha \alpha' \equiv 1(N)$. We observe

$$(y + a_n x_n)^2 \equiv (y - \alpha' \beta x_n)^2 + \frac{2\alpha'}{N}(\alpha a_n + \beta)^2 \pmod{2\mathbb{Z}}$$

Thus

(3.3) $$\sum_{a_n=0}^{N-1} e^{\pi i k(y + a_n x_n)^2} = e^{\pi i k(y - \alpha' \beta x_n)^2} \sum_{a_n=0}^{N-1} e^{\pi i k \frac{2\alpha'}{N}(\alpha a_n + \beta)^2}.$$

The second factor on the right hand side of (3.3) can be calculated explicitly. As $\alpha' \ne 0$, we find for $N \equiv 3(4)$

(3.4a) $$= \sum_{r=0}^{N-1} e^{\frac{2\pi i}{N} k \alpha' r^2} = \left(\frac{k\alpha'}{N}\right) \sqrt{N} i = \left(\frac{k\alpha}{N}\right) \sqrt{N} i$$

and for $N \equiv 1(4)$

(3.4b) $$= \sum_{r=0}^{N-1} e^{\frac{2\pi i}{N} k \alpha' r^2} = \left(\frac{k\alpha'}{N}\right) \sqrt{N} = \left(\frac{k\alpha}{N}\right) \sqrt{N}.$$

Now consider the first factor. Let $\phi(y) = y - (y, x_n) N \alpha' x_n$. The set

$$\left\{ \phi(y) \mid y = \sum_{i=1}^{n-1} a_i x_i, 0 \le a_i \le N - 1 \right\}$$

forms an abelian group on $n - 1$ generators. It has exactly the $n - 1$ relations inherited from L^*/L. Hence we evaluate (3.2) using equations (3.4) and induction on the number of generators as in equation (3.3). We obtain for $N \equiv 3(4)$

$$c_0(k) = \frac{e^{-3l\pi i/4}}{\sqrt{|L^*/L|}} \prod_{r=1}^{n} \left(i \left(\frac{k\alpha_r}{N}\right) \sqrt{N} \right) = e^{n\pi i/2 - 3l\pi i/4} \left(\frac{k^n}{N}\right) \left(\frac{\prod_{r=1}^{n} \alpha_r}{N}\right)$$

3.2. THE CHARACTER OF THETA

For $N \equiv 1(4)$ we obtain

$$c_0(k) = \frac{e^{-3l\pi i/4}}{\sqrt{|L^*/L|}} \prod_{r=1}^{n} \left(\left(\frac{k\alpha_r}{N}\right)\sqrt{N}\right) = e^{-3l\pi i/4}\left(\frac{k^n}{N}\right)\left(\frac{\prod_{r=1}^{n}\alpha_r}{N}\right)$$

Next we observe that $(V_1, *) = (S, +)(T, 1)(S, +)(T, 1)(S, +) = (T^{-1}, 1)$ acts trivially, i.e. $c_0(1) = 1$. Hence, for general k, and any prime $N \neq 2$, $c_0(k) = \left(\frac{k}{N}\right)^n$.

Because the coefficient $c_0(k)$ has modulus 1 and because the transformation induced by the action of V_k is unitary (see the final part of theorem 3.3), the remaining coefficients must be zero. The Legendre symbol is multiplicative and thus we have proven

THEOREM 3.4. *Let L be even l-dimensional integral lattice, let L^*/L have n generators, let $N \neq 2$ be prime such that $NL^* \subseteq L$. Then θ_L is a modular form for $\Gamma_0(N)$ of weight $\frac{l}{2}$ with character $\chi\binom{ab}{cd} = \left(\frac{d}{N}\right)^n$. (The character of course depends on the branch and is as stated when the branch of any $(G, *)$ is the product of the branches of the $(V_k, *)$ above.)* □

We observe two interesting consequences of the above argument. The fact that the action of $(V_1, 1)$ is trivial provides us with some relations between l, n and N. In the case $N \equiv 3(4)$ we conclude that $n \equiv \frac{l}{2} \mod 2$. Thus $\frac{l}{2}$ can replace n in theorem 3.4. In the case $N \equiv 1(4)$ we conclude that $\frac{l}{2}$ must be even.

For $N = 2$ we recall that $\Gamma_0(2)$ was generated by T and ST^2S. The following theorem can be proved along the same lines as theorem 3.4.

THEOREM 3.5. *Let L, l, N, n be as in theorem 3.4, except that we now assume $N = 2$. Then*

$$\theta_L\big|_{\left((S,+)(T^2,1)(S,+),\frac{l}{2}\right)}(q) = e^{-l\frac{\pi i}{2}}\theta_L(q).$$

□

CHAPTER 4

The Proof of Theorem 1.7

In this chapter we prove formula (1.25). This will then complete the proof of the central result of chapter 1. In section 4.1, we describe the lattice $L = \Lambda^{\sigma\perp}$ and its dual L^*. Note that in this chapter, as in chapter 3, L does not denote the same object as in chapter 1. The explicit identification of the fixed point lattices can then be used to complete the proof of lemma 1.3. We identify the modularity properties of the theta-function of L and some lattice symmetries. In section 4.2, we proceed to count the number of elements of L^*/L according to norm. This amounts to counting sums of quadratic residues. Section 4.3 classifies the vectors of L of norm less than 2. Section 4.4 puts everything together to conclude the proof of the central theorem 1.7.

We start with the observation that for $N = 2$ and $N = 3$ the fixed point lattices Λ^σ are well known. For $N = 2$, [**CS88**] identify the fixed point lattice in chapter 4, section 10, as the Barnes-Wall lattice Λ_{16}. Its orthogonal complement is the lattice $\sqrt{2}E_8$. For $N = 3$, [**CS88**] identify the fixed point lattice in chapter 4, section 9, as the Coxeter-Todd lattice K_{12} whose orthogonal complement is K_{12}. Thus in these cases the theta-functions are well known.

4.1. The Theta-Function of L*

[**CS88**], chapter 10, theorem 25 provides the following description of the Leech lattice Λ in $\mathbb{R}^{24}$. Let $\mathcal{C}$ be the set of the elements of the 24-dimensional Golay code. The vector $(x_\infty, x_0, ..., x_{22}) \in \mathbb{Z}^{24}$ is in $\sqrt{8}\Lambda$ if and only if

(4.1a)   the co-ordinates x_i are all congruent modulo 2, to m, say;

(4.1b)   the set of i for which x_i takes any given value modulo 4 is a $\mathcal{C}$-set;

(4.1c)   the co-ordinate-sum is congruent to $4m$ modulo 8.

Equivalently we recall from [**CS88**], chapter 4 that any element of $\sqrt{8}\Lambda$ is of one of the following two shapes:

(4.2) $$2c + 4x \quad \text{or} \quad (1^{(24)}) + 2c + 4y$$

Here $c \in \mathcal{C}$ is understood as an element of $\mathbb{R}^{24}$, $(x_n) \in \mathbb{Z}^{24}$ is such that $\sum x_n$ even, $y = (y_n) \in \mathbb{Z}^{24}$ is such that $\sum y_n$ odd. In line with equation (1.3), the norm of an actual Leech lattice element in the above descriptions is $\frac{1}{8}$ times the sum of the squares of the (integer) co-ordinates.

We consider simultaneously the cases $N = 2, 3, 5, 7, 11, 23$. Let σ be an automorphism of the Leech lattice of order N and cycle shape $1^M N^M$ where $M = \frac{24}{N+1}$. Thus σ acts by permuting the co-ordinates of $\mathbb{R}^{24}$. The character of the 24-dimensional representation (where σ acts as permutation of the co-ordinates) is registered in the Atlas [**Con85**] as χ_{102}. From the cycle shape of σ we conclude

4.1. THE THETA-FUNCTION OF L*

that $\chi_{102}(\sigma) = M$. Hence we can identify the conjugacy class of σ in the Atlas [**Con85**] as 2A, 3B, 5B, 7B, 11A, and 23A respectively.

In the case $N = 5$ we observe that the five elements of a cycle of σ exactly determine an octad. The action of σ maps this into another octad as σ preserves the Leech lattice as a whole. This new octad also contains the 5-cycle, hence the octad is preserved by σ. Thus the 3 remaining points of the octad must be among the 4 fixed points of σ. A direct check of the examples given in [**CS88**], chapter 10, section 2.1, yields that every 7-cycle together with one of the fixed points forms an octad of $\mathcal{C}$ and that every 11-cycle together with one of the fixed points forms a dodecad of $\mathcal{C}$.

To determine the elements of Λ^σ we observe that $\lambda \in \sqrt{8}\Lambda$ is invariant under σ if and only if $\lambda = (a_1, ..., a_M, b_1^{(N)}, ..., b_M^{(N)})$. ($b^{(N)}$ denotes N entries of b in the positions of a cycle of σ. We may order the co-ordinates according to the cycles of σ.) Thus the fixed point lattice Λ^σ is a $2M$-dimensional lattice. We define $L = (\Lambda^\sigma)^\perp$, the lattice orthogonal to the fixed point lattice of dimension $24 - 2M$. By theorem 3.2, L^*/L and $(\Lambda^\sigma)^*/\Lambda^\sigma$ are isomorphic. By theorem 3.1, $(\Lambda^\sigma)^* = \pi\Lambda$, where $\pi = \pi_{\Lambda^\sigma}$ is the projection of the Leech lattice into the subspace spanned by the fixed point lattice.

LEMMA 4.1. *If an element $\lambda = (a_1, ..., b_M^{(N)})$ of $\sqrt{8}(\Lambda^\sigma)^*$ has integer co-ordinates, it is in $\sqrt{8}\Lambda^\sigma$.*

PROOF. For $N = 2, 3$ this follows from the explicit description of the lattices. For the remaining N we observe that π acts by averaging over the cycles. Hence it follows from the description in (4.2) that the co-ordinates of λ are either all even or all odd. We subtract a suitable multiple of $(-3^{(1)}, 1^{(23)}) \in \sqrt{8}\Lambda^\sigma$ to obtain even co-ordinates everywhere. From the description of the cycles of σ with respect to the Golay code we conclude that for every cycle there exists an octad (or dodecad respectively) with entries precisely in one cycle of σ and in a number of co-ordinates fixed by σ. Subtracting suitable multiples of these we obtain from λ an element $\lambda' \in \sqrt{8}(\Lambda^\sigma)^*$ with non-zero entries only in the M co-ordinates fixed by σ. This, in turn, implies that λ' must be an element of $\sqrt{8}\Lambda$ by the following argument. The product of λ' with any element of $\sqrt{8}L$ must be in $8\mathbb{Z}$. λ' has non-zero entries only in up to four places. Now there are elements of $\sqrt{8}L$ having entries $(2, 0, 0, 0, *, ..., *)$. Thus all the entries of λ' must be divisible by 4. Furthermore, there are elements of shape $(1, 1, 1, 1, *, ..., *)$ of $\sqrt{8}L$. Hence the sum of the entries of λ' must be divisible by 8. Thus λ' satisfies the conditions (4.1) and hence is an element of $\sqrt{8}\Lambda$. □

The projection π acts by averaging the co-ordinates of the cycles of the permutation. Thus the entries must be in $\frac{1}{N}\mathbb{Z}$. Hence we have shown that $N(\Lambda^\sigma)^* \subset \Lambda^\sigma$ and more precisely $(\Lambda^\sigma)^*/(\Lambda^\sigma)$ is an M-dimensional vector space over $\mathbb{Z}/N\mathbb{Z}$ and has N^M elements. Moreover, we can give an explicit basis for $(\Lambda^\sigma)^*/\Lambda^\sigma$ as follows:

$$\lambda_j^{\sigma*} = \frac{1}{\sqrt{8}}(4, 0, \ldots, 0, \overset{(N)}{\frac{4}{N}}, 0, \ldots)$$

where $j = 1, \ldots, M$, one for each N-cycle of σ. Note that the $\lambda_j^{\sigma*}$ are mutually orthogonal modulo $2\mathbb{Z}$ and of norm $\frac{2}{N}(2\mathbb{Z})$.

By theorem 3.2, the same holds for L^*/L. For the transition we observe that any $\lambda \in \Lambda$ can be written as

$$\lambda = \pi_L(\lambda) + \pi_{\Lambda^\sigma}(\lambda).$$

In particular, this implies that $\pi_L(\lambda) \equiv \pi_L(\mu) \pmod{L}$ if and only if $\pi_{\Lambda^\sigma}(\lambda) \equiv \pi_{\Lambda^\sigma}(\mu) \pmod{\Lambda^\sigma}$ and further

$$(\pi_L(\lambda), \pi_L(\mu)) \equiv -(\pi_{\Lambda^\sigma}(\lambda), \pi_{\Lambda^\sigma}(\mu)) \bmod \mathbb{Z}$$

L is even because Λ is and L has dimension $24 - 2M$. We can now apply theorems 3.4 and 3.5 to obtain

THEOREM 4.1. *For all $N = 2, 3, 5, 7, 11, 23$, the theta-function of L, θ_L, is a modular form of $\Gamma_0(N)$ of weight $\frac{24-2M}{2} = 12\frac{N-1}{N+1}$. The character is $\chi\binom{ab}{cd} = \left(\frac{d}{N}\right)$ for $N = 7$ and $N = 23$. The character is trivial in the other cases.* □

Using the explicit identification of the dual lattice elements, we are now able to complete the

PROOF OF LEMMA 1.3. The cases $N = 2$ and $N = 3$ are clear as the fixed point lattices are Λ_{16} and K_{12} respectively. So we assume $N \geq 5$ and $M \leq 4$. Now suppose there were a root $r \in \Lambda^\sigma$. The reflection induced by r is the same as that of $nr \in \Lambda^\sigma$ where n is any nonzero integer. Hence we can assume that r is primitive in the sense that, if $n \in \mathbb{Z}, |n| > 1$, then $\frac{r}{n}$ is not in Λ^σ. For any $v \in \Lambda^\sigma$, the inner product (r, v) is integer, hence the greatest common divisor $d = (r, \Lambda^\sigma)$ of all the inner products $(r, v), v \in \Lambda^\sigma$, is defined. It follows that $\frac{r}{d} \in \Lambda^{\sigma*}$. We have seen above that $N\Lambda^{\sigma*} \subset \Lambda^\sigma$. As N is prime it follows that the only possible cases are $d = 1$ or $d = N$.

The reflection through r takes a vector v to the vector $v - \frac{2(r,v)}{(r,r)}r$. If $d = 1$ we conclude that $\frac{2}{(r,r)}$ must be integer, which is impossible as the Leech lattice does not contain elements of norm 1 or norm 2. If $d = N$ we conclude that $\frac{2N}{(r,r)}$ must be integer. This is equivalent to $\frac{2}{N}$ being an integer multiple of $(\frac{r}{N}, \frac{r}{N})$. Now $\frac{r}{N} \in \Lambda^{\sigma*} = \pi_{\Lambda^\sigma}(\Lambda)$. As identified above, elements of $\Lambda^{\sigma*}$ are of shape

$$\frac{1}{\sqrt{8}}(a_1, \ldots, a_M, \frac{b_1}{N}^{(N)}, \ldots, \frac{b_M}{N}^{(N)}).$$

Here $a_i, b_i \in \mathbb{Z}$, and the raised $^{(N)}$ indicates N equal entries. The norm of the above element is

$$\frac{1}{8}(a_1^2 + \cdots + a_M^2 + \frac{b_1^2}{N} + \cdots + \frac{b_M^2}{N}).$$

It remains to prove that this cannot be smaller or equal $\frac{2}{N}$ in the relevant cases.

The projection π from Λ to Λ^σ acts by averaging the entries of the N-cycles. Equation (4.2) stated that the vectors in Λ are of two different shapes. For vectors of shape $(1^{(24)} + 2c + 4y)$ we note that the vector $(1^{(24)})$ is preserved under π. Hence

$$\|\pi(1^{(24)} + 2c + 4y)\| \geq \frac{1}{8}(M \times 1 + M \times \frac{1}{N}) = \frac{1}{8}\frac{M(N+1)}{N} = \frac{3}{N}.$$

For vectors of the remaining shape, $(2c + 4x)$ where c is an element of the Golay-code and the sum of the entries of x is even we observe that the norm will be less or equal $\frac{2}{N}$ only if all a_i equal 0 and at most two of the b_i have modulus at most 1. However, if the sum of coordinates of a cycle is nonzero, it is at least 2, because all entries are even. Thus there cannot be any vectors of norm less or equal $\frac{2}{N}$. □

The group of all automorphisms of the Leech lattice is identified as $2.Co_1$ in the Atlas [**Con85**]. If σ is an automorphism as above we consider the group $\langle\sigma\rangle$ of order N, generated by σ. The normalizer of σ in $2.Co_1$ is the subgroup of $2.Co_1$

of all elements ψ such that $\psi\langle\sigma\rangle = \langle\sigma\rangle\psi$. If ψ is an element of the normalizer of σ it maps the fixed point lattice of σ to itself and hence induces an automorphism of L and moreover L^*. For the same reason it induces an automorphism on L^*/L. For N prime, we write $\mathbb{Z}_N$ for the finite field $\mathbb{Z}/N\mathbb{Z}$. We follow the notation of the Atlas [**Con85**] for orthogonal groups over finite fields. For even dimension $2m$, the '+' type of an orthogonal group corresponds to maximum Witt index m, and the '−' type corresponds to Witt index $m-1$.

CLAIM. *For $N = 11, 7, 5$, and $M = 24/(N+1)$, consider σ of order N as above. For $N = 11$, the centralizer of σ in $2.Co_1$ acts on L^*/L as the orthogonal group $O_2^-(11)$ (up to a conjugacy class of order 2). For $N = 7$ and $N = 5$, the normalizers of σ in $2.Co_1$ act on L^*/L as the orthogonal groups $SO_3(7)$ and $GO_4^+(5)$, respectively.*

PROOF. All calculations required to prove this claim were carried out during the composition of the Atlases [**Con85**], [**Jan95**], but not all have been documented. Theorem 3.2 applies such that we obtain $L^*/L = (\Lambda^{\sigma\perp\,*})/(\Lambda^{\sigma\perp}) \stackrel{\sim}{=} \Lambda^{\sigma\,*}/\Lambda^\sigma$. Hence, we may consider the fixed point lattice, rather than its orthogonal complement. This reduces the dimension of the lattices under consideration from $24-2M$ to $2M$, over $\mathbb{C}$. Furthermore, explicit descriptions of the lattice elements are far more readily available, and a basis has been identified on p. 39 above.

For the case $N = 11$, the element 11A has centralizer $11 \times D_{12}$ in $2.Co_1$. In accordance with the notations of the Atlas [**Con85**], D_{12} here denotes the dihedral group of order 12. We note the isomorphism $D_{12} \stackrel{\sim}{=} O_2^-(11)$. The action of the centralizer on the quotient space $\Lambda^{\sigma\,*}/\Lambda^\sigma$ induces a representation of the orthogonal group over $\mathbb{Z}_{11}$ of degree 2.

Using the Suzuki construction of the Leech lattice we find that the order 3 automorphism within the centralizer corresponds to a rotation through 120 degrees in that construction. Hence its trace must be -12. The character of the 24-dimensional representation of $2.Co_1$ is recorded in the Atlas [**Con85**] as χ_{102}. This identifies the automorphism class as 3A.

When restricted to the action of D_{12}, the 24-dimensional representation of $2.Co_1$ decomposes into a number of irreducible constituents. These are sufficiently characterised by the trace of the automorphism of order 3. Counting multiplicities, there are 12 irreducible constituents whose characters all satisfy

$$\chi(1A) = 2, \quad \chi(2A) = 0, \quad \chi(3A) = -1, \ldots$$

This, in turn, enables us to identify the explicit matrix form of the order 3 automorphism on the 4-dimensional vector space Λ^σ over $\mathbb{C}$, which has trace -2. It is given in formula (6.14) below.

We now consider representations over the finite field $\mathbb{Z}_{11}$. These are in one-to-one correspondence with the representations over $\mathbb{C}$ because the characteristic 11 of $\mathbb{Z}_{11}$ does not divide the group order 12 of D_{12}. Using the explicit basis for $\Lambda^{\sigma\,*}/\Lambda^\sigma$, we find that the automorphism of class 3A continues to act non-trivially on the 2-dimensional space $\Lambda^{\sigma\,*}/\Lambda^\sigma$. Its matrix form over the finite field $\mathbb{Z}_{11}$ is

$$\begin{pmatrix} 5 & -3 \\ 3 & 5 \end{pmatrix} \mod 11.$$

The character table of D_{12} now sufficiently characterises the induced action of D_{12} for our purposes. Note that we have not determined the action of one conjugacy class of order 2 which may, according to character table, act as either $+\text{id}$ or $-\text{id}$. Up to this class however, we may nevertheless conclude that the induced action of the centralizer is the orthogonal matrix action of the orthogonal group $D_{12} \cong O_2^-(11)$.

For the case $N = 7$, the element 7B in Co_1 has normalizer $(7.3 \times L_2(7)).2$, which contains $L_2(7).2 \cong SO_3(7)$, and centralizer $7 \times L_2(7)$, which contains $L_2(7) \cong O_3(7)$. Note that, in the case $N = 7$, we may consider the normalizer and centralizer in Co_1 because the double cover of $L_2(7)$ in $2.Co_1$ is not proper. The action of the normalizer on the quotient space $\Lambda^{\sigma*}/\Lambda^\sigma$ induces a representation of the orthogonal group of degree 3 over $\mathbb{Z}_7$.

The centralizer contains an element ψ of order 3 which cyclically permutes the 3 octads underlying the automorphism σ. Hence, $\text{tr}(\psi) = 0$. Using the explicit basis of $\Lambda^{\sigma*}/\Lambda^\sigma$, we find that ψ induces an automorphism of the 3-dimensional quotient space (over $\mathbb{Z}_7$), again with trace 0.

The Brauer character table of the (unextended) group $L_3(2) \cong L_2(7)$ in the Atlas [**Jan95**] shows that this trace will only be matched by the character of the irreducible representation ϕ_2, which, at the same time, is the character of the orthogonal representation of $SO_3(7)$. Hence, we have uniquely identified the induced action of the normalizer as the orthogonal matrix action of the orthogonal group $SO_3(7)$.

We turn to the case $N = 5$. The normalizer of an element of class 5B in Co_1 is identified in the Atlas [**Con85**] as $(D_{10} \times (A_5 \times A_5).2).2$. In $2.Co_1$, the normalizer of an element of class 5B contains the orthogonal group $GO_4^+(5) \cong 2.O_4^+(5).2^2$ where $O_4^+(5) \cong A_5 \times A_5$. The centralizer of class 5B in $2.Co_1$ is $5 \times 2.(A_5 \times A_5).2$. Unlike $N = 7$, in the case $N = 5$ we must consider the normalizer and centralizer in the double cover $2.Co_1$. The representations involved do not restrict to representations of subgroups of Co_1. The action of the normalizer on the quotient space $\Lambda^{\sigma*}/\Lambda^\sigma$ induces a representation of the orthogonal group $GO_4^+(5)$ of degree 4 over $\mathbb{Z}_5$.

The leading '5×' corresponds to the element 5B itself. In terms of representations, the trailing '.2' relates to split or fused representations. For our purposes, it is therefore sufficient to consider the irreducible representations of the group $2.(A_5 \times A_5)$. Over both $\mathbb{C}$ and the finite field $\mathbb{Z}_5$, these can be constructed from the representations of the group $(2.A_5) \times (2.A_5)$, by quotienting out the kernel of the representation (which corresponds to quotienting out the central 2, identifying the two leading '2.'s). The characters of all representations of $2.A_5$ are listed in the Atlases [**Con85**] and [**Jan95**].

Again, it is easy to identify explicitly a number of elements of the centralizer. There are permutations of the 4 octads underlying the automorphism σ of order 2, 3, 4. There is also the order 3 automorphism which corresponds to a rotation through 120 degrees in the Suzuki construction. The explicit form of these elements of the centralizer is sufficient to conclude that neither of the two subgroups A_5 acts trivially on any 1-dimensional subspace of the 8-dimensional space spanned by the fixed point lattice. Equally, this holds for the 4-dimensional quotient space L^*/L (over $\mathbb{Z}_5$).

This sufficiently characterises the representation in the Atlas character tables of $2.A_5$. Over $\mathbb{C}$, we obtain the 8-dimensional real representation of $2.A_5 \times 2.A_5$ by tensoring the 2-dimensional (conjugate) representations χ_6 or χ_7 and fusing with their algebraic conjugate: $8 = 2 \times 2 + 2 \times 2$. Over $\mathbb{Z}_5$, we obtain the 4-dimensional representation of $2.A_5 \times 2.A_5$ uniquely by tensoring the 2-dimensional representation ϕ_4 such that $4 = 2 \times 2$. At the same time, this is the character of the orthogonal representation of $2.O_4^+(5) \cong 2.(A_5 \times A_5)$. (Note that the 'generically simple' group $O_4^+(5) \cong A_5 \times A_5$ has a faithful *projective* 4-dimensional representation only.) Hence, we have uniquely identified the induced action of the normalizer as the orthogonal matrix action of the orthogonal group $GO_4^+(5)$. □

THEOREM 4.2. *Let N be any of 23, 11, 7, 5, 3, or 2. Let σ of order N and $L = \Lambda^{\sigma\perp}$ be as above. Let λ^*, μ^* be elements of L^*, not in L, such that $\frac{(\lambda^*)^2}{2} \equiv \frac{(\mu^*)^2}{2}$ modulo $\mathbb{Z}$. Then $\theta_{L+\lambda^*} = \theta_{L+\mu^*}$.*

PROOF. Let $[\cdot]$ denote the residue class in L^*/L. It suffices to construct an automorphism of L^* so that its restriction to L^*/L takes the residue class $[\lambda^*]$ of λ^* to the residue class $[\mu^*]$ of μ^*. Now L^*/L is a vector space over the finite field $\mathbb{Z}_N$ of dimension M. We recall the inclusion $N(\Lambda^\sigma)^* \subset \Lambda^\sigma$. Hence, for $N \neq 2$ we obtain a non-degenerate inner product with values in $\mathbb{Z}_N$ if we define

$$([\lambda^*], [\mu^*]) = N(\lambda^*, \mu^*) \bmod N$$

where the bilinear form on the right hand side is the standard form. Calculating the Witt index shows that this form is of '−' type for $N = 11$, and of '+' type for $N = 5$. (For $N = 23$, the form is trivial, for $N = 7$ it is unique. The cases $N = 2$ and $N = 3$ are handled separately.) We will now consider the various N individually. The claim is trivial for $N = 23$ as $M = 1$.

Case $N = 11$: The 121 elements of the 2-dimensional vector space over $\mathbb{Z}_{11}$ fall into 10 classes (of norms 1 to 10, respectively) with 12 elements each, plus the zero vector. Co-ordinate interchange and application of $-\mathrm{id}$ are obvious symmetries. It is straightforward to check that the automorphism $\begin{pmatrix} 5 & -3 \\ 3 & 5 \end{pmatrix}$ mod 11 of order 3 explicitly identified above completes the required symmetries.

Cases $N = 7$ and $N = 5$: We found above that the group $SO_3(7)$, and hence $GO_3(7)$, describes symmetries of the quotient lattice $L^*/L = (\Lambda^{\sigma\perp^*})/(\Lambda^{\sigma\perp})$ for $N = 7$. Equally we found that $GO_4^+(5)$ describes symmetries of the quotient lattice for $N = 5$. Witt's Extension Theorem ([**Jac85**], Chapter 6.5) asserts that, given any non-singular bilinear form (i.e. '+' and '−' type for even dimensions, unique for odd dimensions), any isometry of subspaces (i.e. the map taking $[\lambda^*]$ to $[\mu^*]$) can be extended to an isometry of the whole vector space, i.e. an element of the group $GO_M^\epsilon(N)$.

Case $N = 3$: [**CS83**] prove that the group of isometries of the Coxeter-Todd lattice acts transitively on vectors of norms 4, 6, and 8 respectively, which is sufficient to prove the claim as $L^* = \frac{1}{\sqrt{3}} K_{12}$.

Case $N = 2$: In [**CS88**], chapter 4, section 10 (p.131) the centralizer of σ is identified as $O_8^+(2)$. $O_8^+(2)$ acts transitively on vectors of norm 1 and 2 in $L^* = \frac{1}{\sqrt{2}} E_8$. □

We summarize: The translated theta-function $\theta_{L+\lambda^*}$, $\lambda^* \notin L$, does only depend on $\frac{(\lambda^*)^2}{2}$ mod $\mathbb{Z}$. Hence we can pick λ_r^* such that $\frac{(\lambda_r^*)^2}{2} \equiv \frac{-r}{N}$ mod $\mathbb{Z}$ for $r = 0, \ldots, N-1$ and write

$$(4.3) \qquad \theta_{L^*} = \theta_L + \sum_{r=0}^{N-1} \rho_M(r, N) \theta_{L+\lambda_r^*}.$$

Here $\rho_M(r, N)$ is the number of non-zero elements with half-norm identical to $\frac{-r}{N}$ in L^*/L. It will be determined in the next section.

4.2. Sums of Quadratic Residues

Let N and $M = \frac{24}{N+1}$ be as above. For $\lambda \in \Lambda$ we calculate the norm of $\pi_L \lambda \in L^*$ as

$$(4.4) \qquad (\pi_L \lambda)^2 = \left((1 - \pi_{\Lambda^\sigma})\lambda\right)^2 = \lambda^2 - \left(\pi_{\Lambda^\sigma} \lambda\right)^2.$$

Hence the number $\rho_M(r, N)$ of residue classes of a certain half-norm $\frac{-r}{N} \mod \mathbb{Z}$ in L^*/L is equal to the number of residue classes of half-norm $\frac{r}{N} \mod \mathbb{Z}$ in $\Lambda^{\sigma*}/\Lambda^\sigma$. We recall that we found an explicit basis $\{\lambda_j^{\sigma*}\}$ of $\Lambda^{\sigma*}/\Lambda^\sigma$ such that $\frac{(\lambda_j^{\sigma*})^2}{2} \equiv \frac{1}{N} \mod \mathbb{Z}$ and $\frac{(\lambda_j^{\sigma*}, \lambda_k^{\sigma*})}{2} \equiv 0 \mod \mathbb{Z}$. Thus $\rho_M(r, N)$ is equal to the number of non-zero solutions to the following equation:

$$(4.5) \qquad \sum_{i=1}^{M} x_i^2 \equiv r \mod N$$

for $x_i \in \mathbb{Z}_N$. We now define $\tilde{\rho}_M(r, N)$ to be the number of all solutions to the above equation. Hence for $r \not\equiv 0(N)$, $\tilde{\rho}_M(r, N) = \rho_M(r, N)$, and $\tilde{\rho}_M(0, N) = \rho_M(0, N) + 1$. We observe that $\tilde{\rho}_M(m^2 r, N) = \tilde{\rho}_M(r, N)$ for any non-zero m. Note that we always consider ordered pairs, or n-tuples, of residues.

LEMMA 4.2. *Let $N \neq 2$ be a prime. If $N \equiv 1(4)$ (that is, -1 is a square modulo N) then $\tilde{\rho}_2(0, N) = 2N - 1$ and $\tilde{\rho}_2(r, N) = N - 1$ for all $r \not\equiv 0(N)$. If $N \equiv 3(4)$ (that is, -1 is a non-square modulo N) then $\tilde{\rho}_2(0, N) = 1$ and $\tilde{\rho}_2(r, N) = N + 1$ for all $r \not\equiv 0(N)$.*

PROOF. If -1 is a square, say $z^2 \equiv -1$ then $x^2 + y^2 \equiv 0$ has the following $2N - 1$ solutions $(x, zx), (x, -zx)$ for non-zero x and $(0, 0)$. (note that $z \not\equiv -z$ as $N \neq 2$.) $x^2 + y^2 \equiv r$ is equivalent to $x^2 - (zy)^2 = (x + zy)(x - zy) \equiv r$. This clearly has the same number of solutions for any non-zero r. The result follows as there are $N^2 - 2N + 1$ possible choices of (x, y) and $N - 1$ different non-zero values for r.

If -1 is a non-square modulo N, $\tilde{\rho}_2(0, N) = 1$ obviously. Let us first consider $\tilde{\rho}_2(r, N)$ for a square $r = z^2$. The equation $x^2 + y^2 \equiv z^2 \mod N$ is equivalent to $x^2 \equiv (z + y)(z - y) \mod N$. We substitute $a = z + y$ and $b = z - y$. Now if we choose $a = 0$ then b is arbitrary and $x = 0$. (Note that we consider free z.) This gives N solutions. If we choose $a \neq 0$ ($N - 1$ choices) then we can furthermore choose x arbitrarily which fixes b. This gives $(N - 1)N$ solutions. We exclude the solution $(x, y, z) = (0, 0, 0)$ and are left with $N^2 - 1$ solutions such that $z \neq 0$. Of course, the number of solutions for any of the $N - 1$ non-zero z must be the same, hence there are $N + 1$ such. This proves $\tilde{\rho}_2(r, N) = N + 1$ for any square r. $\tilde{\rho}_2(r, N) = N + 1$ for r a non-square then follows from the fact that there are $\frac{1}{2}(N - 1)(N + 1)$ pairs (x, y) left and that there are $\frac{1}{2}(N - 1)$ non-squares. $\square$

COROLLARY. *For M even, $\tilde{\rho}_M(r, N)$ is equal for any non-zero r.*

PROOF. We obtain the result by induction using

$$\tilde{\rho}_M(r, N) = \sum_{s=0}^{N-1} \tilde{\rho}_2(s, N) \, \tilde{\rho}_{M-2}(r - s, N).$$

$\square$

4.2. SUMS OF QUADRATIC RESIDUES

THEOREM 4.3. *For even M, that is $N = 2, 3, 5, 11$, we have*
$$\tilde{\rho}_M(r \neq 0, N) = \frac{1}{N}\left(N^M - (-N)^{\frac{M}{2}}\right) \text{ and}$$
$$\tilde{\rho}_M(r = 0, N) = N^{M-1} + (-1)^{\frac{M}{2}}\left(N^{\frac{M}{2}} - N^{\frac{M}{2}-1}\right).$$
For odd M, that is $N = 7, 23$, we have
$$\tilde{\rho}_M(r \neq 0, N) = \frac{1}{N}\left(N^M + \left(\frac{r}{N}\right) N^{\frac{M+1}{2}}\right) \text{ and } \tilde{\rho}_M(r = 0, N) = N^{M-1}.$$
Here, $\left(\frac{r}{N}\right)$ stands for the Legendre symbol as before.

PROOF. We use lemma 4.2 to proceed inductively to the cases $\tilde{\rho}_M(r, N)$ of interest. The case $(N, M) = (23, 1)$ is trivial as any square can be represented in 2 ways and any non-square cannot be represented. The case $(N, M) = (11, 2)$ is covered by lemma 4.2.

Case $(N, M) = (7, 3)$:
$$\tilde{\rho}_3(0, 7) = \tilde{\rho}_1(0, 7)\,\tilde{\rho}_2(0, 7) + \sum_{s=1}^{\frac{N-1}{2}} \tilde{\rho}_1(s^2, 7)\,\tilde{\rho}_2(-s^2, 7)$$
$$= 1 \times 1 + \frac{(N-1)}{2} \times 2 \times (N+1) = N^2$$
$$\tilde{\rho}_3(-1, 7) = \tilde{\rho}_1(0, 7)\,\tilde{\rho}_2(-1, 7) + \sum_{s=1}^{\frac{N-1}{2}} \tilde{\rho}_1(s^2, 7)\,\tilde{\rho}_2(-1-s^2, 7)$$
$$= 1 \times (N+1) + \frac{(N-1)}{2} \times 2 \times (N+1) = N(N+1)$$

Note that -1 is not a square, hence $-1 - s^2$ is never zero. The value for r a square now follows as the difference to the total N^3.
$$\tilde{\rho}_3(1, 7) = \frac{N^3 - \tilde{\rho}_3(0, 7) - \frac{(N-1)}{2}\tilde{\rho}_3(-1, 7)}{\frac{N-1}{2}} = N(N-1)$$

Case $(N, M) = (5, 4)$:
$$\tilde{\rho}_4(0, 5) = \tilde{\rho}_2(0, 5)\,\tilde{\rho}_2(0, 5) + \sum_{s=1}^{N-1} \tilde{\rho}_2(s, 5)\,\tilde{\rho}_2(-s, 5)$$
$$= (2N-1) \times (2N-1) + (N-1) \times (N-1) \times (N-1)$$
$$= N^3 + N^2 - N$$

As $M = 4$ is even, for non-zero r we obtain
$$\tilde{\rho}_4(r, N) = \frac{N^4 - (N^3 + N^2 - N)}{N - 1} = N^3 - N$$

Case $(N, M) = (3, 6)$: The squares modulo 3 are 0 and 1. The sum of six squares will be zero modulo 3 if and only if 0, 3, or 6 summands equal 1. Hence there are $\binom{6}{0}2^0 + \binom{6}{3}2^3 + \binom{6}{6}2^6 = 225$ solutions. As $M = 6$ is even, for non-zero r we obtain

$$\tilde{\rho}_6(r,3) = \frac{3^6 - 225}{2} = 252.$$

Case $(N, M) = (2, 8)$: The squares modulo 2 are 0 and 1. The sum of eight squares will be zero modulo 2 if and only if an even number of summands equal 1. of 8 squares by either 0,2,4,6,8 summands 1. Hence $\tilde{\rho}_8(0,2) = \binom{8}{0} + \binom{8}{2} + \binom{8}{4} + \binom{8}{6} + \binom{8}{8} = 136$ and $\tilde{\rho}_8(1,2) = 2^8 - 135 = 120$ which completes the proof. □

4.3. The Short Vectors of L*

In order to identify θ_L we will determine the leading coefficients of both θ_L and θ_{L^*}. As L is a lattice of dimension $24\frac{N-1}{N+1}$, by theorems 3.4 and 3.5 its theta-function has weight $12\frac{N-1}{N+1}$. We will want to apply theorem 2.1. Hence we need to determine the terms of a q-expansion of exponent up to and including exponent $\frac{N-1}{N+1}$. L is a sublattice of the Leech lattice. Hence any non-zero vector has norm greater or equal to 4. Thus the expansion of $\theta_L - 1$ has a leading term of exponent at least 2. We now turn to L^*. For the purpose of the proof it is sufficient to determine those vectors of norm less or equal to $2\frac{N-1}{N+1}$ in L^*. In fact, we will find that there are no such non-zero vectors. Note that the techniques developed below can also be applied to determine the shortest non-zero vectors in the lattices L^*, which turn out to be of norm $2\frac{N-1}{N}$. The calculations were carried out as a double check of the results, but will not be presented here.

The projection π_L onto the orthogonal complement of Λ^σ is identical to the composition of the following two operations on vectors of $\mathbb{R}^{24}$.

π_1: Kill the components which correspond to the M fixed points of σ. This is a map $\mathbb{R}^{24} \to \mathbb{R}^{24-M}$.

π_2: For each of the M N-cycles of σ, subtract a suitable (possibly non-integer) multiple of $(0^{(N(M-1))}, 1^{(N)})$ (non-zero entries in the positions of an N-cycle) such that the sum of the co-ordinates of that cycle becomes 0. This maps $\mathbb{R}^{24-M}$ into a $(24 - 2M)$-dimensional subspace.

Recalling the description of the Leech lattice in chapter 4.1, formula (4.2), we note that we can describe the elements of $\sqrt{8}L^*$ as all elements $\pi_L(2c + 4x)$ where c is still in the Golay code, but x now is arbitrary in $\mathbb{Z}^{24}$. In particular, we only need to consider even entries. The following argument will only depend on the allocation of the co-ordinates to the individual cycles and will be independent of the particular order of the co-ordinates. Hence we adopt the convention that vectors will be represented cycle by cycle without specification of the order of the entries within a cycle or of the order of the cycles. A general element λ of $\sqrt{8}\pi_1\Lambda$ will thus be represented as

(4.6) $$\lambda = ((a_{1,1}, ..., a_{1,N}), ..., (a_{M,1}, ..., a_{M,N})) \in \mathbb{Z}^{24-M}.$$

We introduce the average of a cycle $\bar{a}_m = \frac{1}{N}\sum_{n=1}^N a_{mn}$ and proceed to calculate the norm of $\frac{1}{\sqrt{8}}\pi_2(\lambda) \in L^*$.

(4.7) $$\pi_2\lambda = ((a_{1,1} - \bar{a}_1, ...), ...) \in \mathbb{R}^{24-M}.$$

(4.8)
$$\|\pi_2 \frac{1}{\sqrt{8}}\lambda\| = (\pi_2 \frac{1}{\sqrt{8}}\lambda)^2 = \frac{1}{8}\sum_m\sum_n (a_{mn} - \bar{a}_m)^2 = \frac{1}{8}\sum_{m=1}^{M}\left(\sum_{n=1}^{N} a_{mn}^2 - N\bar{a}_m^2\right).$$

Note that, in the notation of formula (4.2), $\frac{1}{\sqrt{8}}\lambda$ is the actual element of L^* and that the norm of an element is understood to be the inner product with itself rather than the square root of this inner product. Formula (4.8) proves the following

LEMMA 4.3. *If λ is as in formula (4.6) above then*
$$\|\pi_2\lambda\| \geq \|\pi_2\left((|a_{1,1}|,...,|a_{1,N}|),...,(|a_{M,1}|,...,|a_{M,N}|)\right)\|.$$

The inequality will be strict if within a cycle there are strictly negative and strictly positive entries. □

Furthermore, to achieve minimal norm the entries must be spread out as evenly as possible in the sense of

LEMMA 4.4. *Let λ be as in formula (4.6) above. Suppose that within a cycle co-ordinate entries differ by more than 2. Suppose further that the entries of the cycle are not all equivalent modulo 4. Then there exists a $\mu \in \sqrt{8}\pi_1\Lambda$ such that $\|\pi_2\lambda\| > \|\pi_2\mu\| > 0$.*

PROOF. We use the notation of formula (4.8). Recall that we only need to consider $a_{mn} \in 2\mathbb{Z}$. For fixed average $\bar{a}_m$ of a cycle the minimum norm is obviously attained if the $a_{mn} \in 2\mathbb{Z}$ have no greater pairwise difference than two. Note that any vector that does not satisfy the condition has a strictly greater norm. Note further that the norm does not depend on the value of the average because of π_2. The vectors $((4, 0^{(N-1)}), (0^{(N)}),...)$ are elements of $\sqrt{8}\pi_1\Lambda$. Hence, if $\lambda \in \sqrt{8}\pi_1\Lambda$ satisfies the assumptions of lemma 4.4 it is possible to subtract suitable multiples of these in order to obtain a vector μ which satisfies the claimed inequality. $\pi_2\mu = 0$ is equivalent to all the co-ordinate entries being equivalent modulo 4. □

LEMMA 4.5. *Let*
(4.9)
$$\lambda_q = \left((4^{(q)}, 0^{(N-q)}), (0^{(N)}),...\right).$$

Then $\|\pi_2\lambda_q\| = 2\frac{q(N-q)}{N}$. There are no vectors with entries equivalent 0 modulo 4 which are mapped by π_2 to non-zero vectors with norm less or equal $2\frac{N-1}{N+1}$.

PROOF. The first claim is a straightforward calculation. Obviously none of the vectors λ_q has norm less or equal $2\frac{N-1}{N+1}$. For the second claim it remains to consider vectors whose entries differ by at least 8. We use the argument of the proof of lemma 4.4. □

The only q with $\|\pi_2\lambda_q\| = 2\frac{N-1}{N}$ are $q = 1$ and $q = N-1$. As $\pi_2\lambda_1 = -\pi_2\lambda_{N-1}$ there are $2(24 - M)$ vectors of the type $\pi_2(\lambda_q)$ and norm $2\frac{N-1}{N}$ in L^*.

Thus we only have to consider vectors of the form $\pi_L(2c)$ where c is an element of the 24-dimensional Golay code, understood as element of $\mathbb{R}^{24}$. We recall the following three properties of the Golay code, for details see [**CS88**]:

(P1) Given a Golay code element d with $\pi_2 d = 0$ we conclude that a Golay code element c has (up to sign) the same image under π_2 as the symmetric difference $c \triangle d$. Note that if c and d meet in, say, a places then $c \triangle d$ and d meet in $|d| - a$ places. This will reduce the number of cases to be considered in the sequel.

(P2) An octad meets a dodecad in 2, 4, or 6, places. Any two distinct octads meet in 0, 2, or 4, places. (Otherwise their symmetric difference would have length not in $\{0, 8, 12, 16, 24\}$.)

(P3) Any dodecad is the symmetric difference of two octads.

We furthermore introduce the following shorthand. For $N = 11$,

$$(4.10) \qquad (a_1, a_2) := \left((0^{(11-a_1)}, 2^{(a_1)}), (0^{(11-a_2)}, 2^{(a_2)}) \right) \in \sqrt{8}\pi_1 \Lambda.$$

Furthermore, given a Golay code element c, we define its intersection numbers (x_1, x_2) with respect to the dodecads d_1 and d_2. An analogous notation will be used for all N.

Case N=23: $\|\pi_2(0^{(23-n)}, 2^{(n)})\| = \frac{1}{2} \frac{n(23-n)}{23}$. This is smaller than 2 only for $n < 6$ or $n > 17$. The 24-dimensional Golay code is such that any element c has 0,8,12,16 or 24 entries 1, the rest being zero. As the projection kills the first coordinate the projected Golay codes can have 0, 7, 8, 11, 12, 15, 16 entries. This completes the analysis of any non-zero Golay code element. Hence there are no non-zero vectors of norm less or equal $2\frac{N-1}{N+1}$.

Case N=11: Let d_1, d_2 denote the two dodecads which define the cycle shape of the automorphism. Given an octad o such that $\pi_1(o) = (a_1, a_2)$ in the notation of (4.10) we obtain the norm as

$$\|\pi_2(a_1, a_2)\| = \frac{1}{22}\bigl(a_1(11 - a_1) + a_2(11 - a_2)\bigr).$$

By (P2), up to permutation there are the following possibilities for the intersection numbers (x_j): (6,2), or (4,4). Then $a_j = x_j$ or $a_j = x_j - 1$ depending on the fact which entries of o are killed by π_1. Note further that $a_j = 6$ and $a_j = 5$ are equivalent by (P1). Hence we need only consider the following cases: $(a_j) = (6,2)$, (6,1), (4,4), (4,3), (3,3). The norms are easily obtained as 24/11, 20/11, 28/11, 26/11, 24/11 respectively. None of these vectors has norm less or equal $2\frac{N-1}{N+1}$.

As dodecads cannot intersect in more than 8 points it is straightforward to see that there cannot arise any vectors of norm less than 2 from a dodecad. The vectors related to 16-point elements of the Golay code have been accounted for by considering octads by (P1).

Because of lemma 4.4 this concludes the proof that there are no non-zero vectors of norm less or equal to $2\frac{N-1}{N+1}$.

Case N=7: Let o_1, o_2, o_3 denote the three octads which define the cycle shape of the automorphism. Given an octad o such that $\pi_1(o) = (a_1, a_2, a_3)$ in the notation of (4.10) we obtain the norm as

$$\|\pi_2(a_1, a_2, a_3)\| = \frac{1}{14}\left(a_1(7 - a_1) + a_2(7 - a_2) + a_3(7 - a_3)\right)$$

Given an octad $o \neq o_j$ of the Golay code it meets o_j in x_j places where $x_j \in \{0, 2, 4\}$ because of (P2). Hence up to permutation (x_j) must be one of the following: (4,4,0), or (4,2,2). Note that $a_j = x_j$ or $a_j = x_j - 1$ (π_1 kills one of the places of the octad o_j). Note further that $a_j = 3$ and $a_j = 4$ are equivalent because of (P1). Hence we only need to consider the following cases for (a_j): (4,4,0), (4,2,2), (4,2,1), (4,1,1).

These have norms 12/7, 16/7, 14/7, 12/7 respectively. There are no cases of norm less or equal to $2\frac{N-1}{N+1}$.

Next we consider a dodecad d. If it intersects with o_j in 6 places then $d - o_j$ is an octad. Hence we have already accounted for it. Otherwise, by (P2) d meets each o_j in 4 places. Hence the projection has norm 18/7. As before, the vectors related to 16-point elements of the Golay code have been accounted for by considering the octads. Because of lemma 4.4 this concludes the list of vectors of norm less or equal to $2\frac{N-1}{N+1}$.

Case N=5: Let o_1, o_2, o_3, o_4 denote the four octads which define the cycle shape of the automorphism. Note that this time the defining octads intersect, though only in places which are killed by π_1. Given an octad o such that $\pi_1(o) = (a_1, a_2, a_3, a_4)$ in the notation of (4.10) we obtain the norm as

$$\|\pi_2((a_j)_{j=1}^4)\| = \frac{1}{10} \sum_{j=1}^4 a_j(5 - a_j).$$

Given an octad $o \neq o_j$ of the Golay code it meets the octads o_j in $x_j, j = 1, ..., 4$ places respectively. Because of (P1) it is sufficient to consider the cases where $a_j \in \{0, 1, 2\}$. The number k of co-ordinates killed by π_1 satisfies $0 \leq k = 8 - \sum a_j \leq 4$. Hence up to permutation (a_j) must be one of the following: (2,2,2,2), (2,2,2,1), (2,2,2,0), (2,2,1,1), (2,2,1,0), (2,2,0,0), (2,1,1,1), (2,1,1,0), (1,1,1,1). If $k = 4$ the intersection numbers x_j of o with o_j are just $x_j = a_j + 3$. Because of (P2) the x_j must be less or equal 4. This excludes (2,2,0,0) and (2,1,1,0). The remaining cases - whether actually occurring or not - have norm greater than $2\frac{N-1}{N+1}$ and need not be considered.

Next we consider a dodecad d. The sum of the intersection numbers a_j with the 5-cycles must be at least 8. If $(a_j) = (2, 2, 2, 2)$ the projection has norm 12/5. Else there are intersection numbers of 3 or higher. By (P1) the symmetric difference with a suitable o_j will be a dodecad or octad with smaller intersection numbers. Hence we have already accounted for it. As before, the vectors related to 16-point elements of the Golay code have been accounted for by considering the octads. Because of lemma 4.4 this concludes the list of vectors of norm less or equal to $2\frac{N-1}{N+1}$.

Case N=3: As established above, $L = K_{12}$. Hence $L^* = \frac{1}{\sqrt{3}} K_{12}$ and we quote from [**CS88**], chapter 4.9 that

$$\theta_{L^*}(q) = 1 + 756 q^{\frac{2}{3}} + \ldots$$

Case N=2: As established above, $L = \sqrt{2} E_8$. Hence $L^* = \frac{1}{\sqrt{2}} E_8$ and we quote from [**CS88**], chapter 4.8 that

$$\theta_{L^*}(q) = 1 + 240 q^{\frac{1}{2}} + \ldots \qquad \square$$

4.4. The Conclusion of the Proof

So far we have established that θ_L is a modular form of $\Gamma_0(N)$ and have determined the leading coefficients of θ_L and θ_{L^*}. We now turn to the right hand side of formula (1.25). For $N = 2, 3, 5, 7, 11, 23$, let $M = \frac{24}{N+1}$ and define for $r = 0, \ldots, N-1$

$$\Theta(\tau) := \eta(\tau)^{MN} \left(\eta(N\tau)^{-M} + \sum_{j=0}^{N-1} \psi_j(\tau)^{-M} \right),$$

$$\Theta_r(\tau) := \eta(\tau)^{MN} \left(\sum_{j=0}^{N-1} e^{-j\frac{-2r}{N}\pi i} \psi_j(\tau)^{-M} \right) = \eta(\tau)^{MN} \left(\sum_{j=0}^{N-1} e^{jr\frac{2\pi i}{N}} \psi_j(\tau)^{-M} \right).$$

(Remember that we chose λ_r^* such that $(\lambda_r^*)^2 \equiv \frac{-2r}{N} \mod 2\mathbb{Z}$.)

THEOREM 4.4. Θ is a modular form of $\Gamma_0(N)$ of weight $12\frac{N-1}{N+1} = \frac{NM-M}{2}$ and character χ_N. For $N = 2, 3, 5, 11$, χ_N is trivial, for $N = 7, 23$, $\chi_N\left(\begin{pmatrix} ab \\ cd \end{pmatrix}, *\right) = \left(\frac{d}{N}\right)$. (Here, as in theorems 3.4 and 3.5, we implicitly understand $*$ to be the branch as obtained when the group element is represented as a product of generators $(T, 1)$ and $(V_k, *)$ of formula (2.1).)

PROOF. Consider $N = 7, 23$. $\Gamma_0(N)$ is generated by T and the set of $V_k = ST^k ST^{k'} S$ (formula (2.1)). Now from the results of chapter 2.2,

$$\Theta|_{(T,1,\frac{NM-M}{2})}(\tau)$$

(4.11)
$$= \eta(\tau)^{NM}|_{(T,1,\frac{NM}{2})} \left(\eta(N\tau)^{-M}|_{(T,1,-\frac{M}{2})} + \sum_{j=0}^{N-1} \psi_j(\tau)^{-M}|_{(T,1,-\frac{M}{2})} \right)$$

$$= e^{\frac{\pi i}{12}NM}\eta(\tau)^{NM} \left(e^{N\frac{\pi i}{12}(-M)}\eta(N\tau)^{-M} + \sum_{j=0}^{N-1} e^{-\frac{\pi i}{12}(-M)}\psi_j(\tau)^{-M} \right)$$

$$= \Theta(\tau).$$

Here we used $NM + M \equiv 0 \pmod{24}$.

(4.12)
$$\Theta|_{(V_k,*,\frac{NM-M}{2})}(\tau)$$
$$= \eta(\tau)^{NM}|_{(V_k,*,\frac{NM}{2})} \left(\eta(N\tau)^{-M}|_{(V_k,*,-\frac{M}{2})} + \sum_{j=0}^{N-1} \psi_j(\tau)^{-M}|_{(V_k,*,-\frac{M}{2})} \right)$$

Now, $(V_k, *, \frac{NM}{2}) = ((S, +)(T^k, 1)(S, +)(T^{k'}, 1)(S, +), \frac{NM}{2})$ acts on $\eta(\tau)^{NM}$ as multiplication by

(4.13)
$$\exp\left(\left(\frac{-3\pi i}{4} + \frac{(k+k')\pi i}{12}\right)NM\right)$$

We now proceed to calculate the action of $(V_k, *, -\frac{M}{2})$ in the second factor of the right hand side of equation (4.12). This will have to be done case by case for all the summands. We adopt the following shorthand: $\phi_1 \overset{G}{\mapsto} \phi_2$ means that the action of G takes ϕ_1 to ϕ_2. The individual actions applied in every step have been calculated in chapter 2.2, in particular theorem 2.3. To simplify notation we work for M rather than $-M$.

4.4. THE CONCLUSION OF THE PROOF

$$\eta(N\tau)^M \xmapsto{(S,+,\frac{M}{2})} \frac{e^{-M\frac{\pi i}{4}}}{\sqrt{N}^M}\psi_0^M$$

$$\xmapsto{(T^k,1,\frac{M}{2})} \frac{e^{-M\frac{\pi i}{4}}}{\sqrt{N}^M}e^{-kM\frac{\pi i}{12}}\psi_k^M$$

$$\xmapsto{(S,+,\frac{M}{2})} \frac{e^{-4M\frac{\pi i}{4}}}{\sqrt{N}^M}\left(\frac{-k}{N}\right)e^{-kM\frac{\pi i}{12}}\psi_{-k'}^M$$

$$\xmapsto{(T^{k'},1,\frac{M}{2})} \frac{e^{-4M\frac{\pi i}{4}}}{\sqrt{N}^M}\left(\frac{-k}{N}\right)e^{(-k-k')M\frac{\pi i}{12}}\psi_0^M$$

$$\xmapsto{(S,+,\frac{M}{2})} e^{-5M\frac{\pi i}{4}}\left(\frac{-k}{N}\right)e^{(-k-k')M\frac{\pi i}{12}}\eta(N\tau)^M$$

For $j \neq 0, k'$ we calculate

$$\psi_j^M \xmapsto{(S,+,\frac{M}{2})} e^{-3M\frac{\pi i}{4}}\left(\frac{-j'}{N}\right)\psi_{-j'}^M$$

$$\xmapsto{(T^k,1,\frac{M}{2})} e^{-3M\frac{\pi i}{4}}\left(\frac{-j'}{N}\right)e^{-kM\frac{\pi i}{12}}\psi_{-j'+k}^M$$

$$\xmapsto{(S,+,\frac{M}{2})} e^{-6M\frac{\pi i}{4}}\left(\frac{-j'}{N}\right)\left(\frac{-(-j'+k)'}{N}\right)e^{-kM\frac{\pi i}{12}}\psi_{-(-j'+k)'}^M$$

$$\xmapsto{(T^{k'},1,\frac{M}{2})} e^{-6M\frac{\pi i}{4}}\left(\frac{-j'}{N}\right)\left(\frac{-(-j'+k)'}{N}\right)e^{(-k-k')M\frac{\pi i}{12}}\psi_{-(-j'+k)'+k'}^M$$

$$\xmapsto{(S,+,\frac{M}{2})} e^{-9M\frac{\pi i}{4}}\left(\frac{-j'}{N}\right)\left(\frac{-(-j'+k)'}{N}\right)\left(\frac{-(-(-j'+k)'+k')'}{N}\right)e^{(-k-k')M\frac{\pi i}{12}} \times$$

$$\times \psi_{-(-(-j'+k)'+k')'}^M$$

$$= e^{-5M\frac{\pi i}{4}}\left(\frac{-k}{N}\right)e^{(-k-k')M\frac{\pi i}{12}}\psi_{-(-(-j'+k)'+k')'}^M.$$

For the last equality we note that $\left(\frac{k}{N}\right) = \left(\frac{k'}{N}\right) = -\left(\frac{-k}{N}\right)$ and that the Legendre symbol is multiplicative. Thus $\left(\frac{-j'}{N}\right)\left(\frac{-(-j'+k)'}{N}\right)\left(\frac{-(-(-j'+k)'+k')'}{N}\right) =$
$\left(\frac{-j}{N}\right)\left(\frac{(-j'+k)}{N}\right)\left(\frac{(-(-j'+k)'+k')}{N}\right) = \left(\frac{-j}{N}\right)\left(\frac{-1+(-j'+k)k'}{N}\right) = \left(\frac{-j}{N}\right)\left(\frac{-j'k'}{N}\right) = \left(\frac{k}{N}\right)$.

Similar calculations yield

$$\psi_0^M \xmapsto{(V_k,*,\frac{NM}{2})} e^{-5M\frac{\pi i}{4}}\left(\frac{-k}{N}\right)e^{(-k-k')M\frac{\pi i}{12}}\psi_{-k}^M,$$

$$\psi_{k'}^M \xmapsto{(V_k,*,\frac{NM}{2})} e^{-5M\frac{\pi i}{4}}\left(\frac{-k}{N}\right)e^{(-k-k')M\frac{\pi i}{12}}\psi_0^M.$$

We combine formula (4.13) with the above four calculations to obtain that

$$(V_k,*,\frac{NM-M}{2}) = ((S,+)(T^k,1)(S,+)(T^{k'},1)(S,+),\frac{NM-M}{2})$$

acts on Θ as multiplication by $e^{(-3N+5)M\pi i/4}\left(\frac{-k}{N}\right) = \left(\frac{-k}{N}\right)$ (remember to revert to $-M$ for M in the second factor!). Combining this with equation (4.11) we have proven the claimed modularity properties of Θ for $N = 7, 23$ for all generators and hence all elements of $\Gamma_0(N)$.

The proof for $N = 3, 5, 11$ is a simplification of the above argument as we use theorem 2.2 instead of theorem 2.3. For $N = 2$, consider the transformation behaviour under $(S,+)(T^2,1)(S,+)$ instead. This concludes the proof of theorem 4.4. □

A comparison with the results of theorem 4.1 then proves

LEMMA 4.6. θ_L and Θ are both modular forms of $\Gamma_0(N)$ of weight $12\frac{N-1}{N+1}$ with identical character. □

Hence all that remains to be checked is whether the first coefficients of $\theta_L|_{(S,+,\frac{NM-M}{2})}$ equal those of $\Theta|_{(S,+,\frac{NM-M}{2})}$. The transformation under S follows from the results of chapter 2.2

$$(-i\sqrt{N})^M \Theta|_{(S,+,\frac{NM-M}{2})}(q) =$$

$$= \eta(q)^{NM} \left(\eta(q^N)^{-M} + N^M \psi_0(q)^{-M} + (i\sqrt{N})^M \sum_{j=1}^{N-1} \left(\frac{j}{N}\right)^M \psi_j(q)^{-M} \right)$$

It is a straightforward evaluation of the above formula to determine that the lowest positive power of q with non-zero coefficient is $\frac{N-1}{N}$ for all N.

THEOREM 4.5. $\theta_L(q) = \Theta(q)$.

PROOF. We recall that the corollary to theorem 2.1 stated that a modular form ϕ of $\Gamma_0(N)$ of weight k vanishes identically if the coefficients of the q-expansions of both ϕ and $\phi|_S$ vanish to order $\frac{k}{12}$. We will apply this theorem to $\phi(q) = \theta_L(q) - \Theta(q)$. Note that it follows from lemma 4.6 above that this is a modular form of $\Gamma_0(N)$ with character of order 2. Furthermore, we obtain $k = 12\frac{N-1}{N+1}$. Thus we need to consider the exponents less than $\frac{N-1}{N+1}$ in the q-expansions of ϕ and $\phi|_S$. These are trivially zero for ϕ itself. In order to calculate the leading coefficients in the q-expansions of $\phi|_{(S,+,\frac{M}{2})} = \Theta|_{(S,+,\frac{M}{2})} - \theta_L|_{(S,+,\frac{M}{2})}$ we note that, by the corollary to theorem 3.3, $(-i)^M \sqrt{N}^M \theta_L|_{(S,+,\frac{M}{2})} = \theta_{L^*}$. The leading coefficients of θ_{L^*} up to and including exponent $\frac{N}{N+1}$ have been determined in chapter 4.3. From those results we conclude that the coefficients of the q-expansion of $\phi|_S$ up to $\frac{N-1}{N+1}$ are zero. □

To complete the proof of theorem 1.7 we have to verify formula (1.25) which can be rewritten as

$$\theta_{L+\lambda_r^*}(q) = \Theta_r(q).$$

We begin by considering the sums

$$\sum_{j=0}^{N-1} e^{jr\frac{2\pi i}{N}} \psi_j(q)^{-M} = \sum_{j=0}^{N-1} e^{jr\frac{2\pi i}{N}} \sum_m c(m) \left(e^{j\frac{2\pi i}{N}} q^{\frac{1}{N}}\right)^m$$

$$= \sum_m \sum_{j=0}^{N-1} \left(e^{\frac{2\pi i}{N}}\right)^{jr+jm} c(m) q^{\frac{m}{N}}.$$

As $\sum_{j=0}^{N-1} e^{-jx\frac{2\pi i}{N}} = 0$ unless $x \equiv 0 \mod N$ it follows that Θ_r has a q-expansion with exponents in $\mathbb{Z} - \frac{r}{N}$ only.

4.4. THE CONCLUSION OF THE PROOF

Furthermore, we calculate for $N = 7, 23$, using $\rho(r) = \rho_M(r, N)$ of theorem 4.3,

$$\Theta + \sum_{r=0}^{N-1} \rho(r)\Theta_r$$

$$= \eta(q)^{NM}\left(\eta(q^N)^{-M} + \sum_{j=0}^{N-1} \psi_j(q)^{-M} + \sum_{r=0}^{N-1} \rho(r)\sum_{j=0}^{N-1} e^{jr\frac{2\pi i}{N}}\psi_j(q)^{-M}\right)$$

$$= \eta(q)^{NM}\left(\eta(q^N)^{-M} + \left(1 + \sum_{r=0}^{N-1} \rho(r)\right)\psi_0(q)^{-M}\right.$$

$$\left. + \sum_{j=1}^{N-1}\left(1 + \sum_{r=0}^{N-1} \rho(r)e^{jr\frac{2\pi i}{N}}\right)\psi_j(q)^{-M}\right)$$

Now $1 + \sum_{r=0}^{N-1} \rho(r)$ is the number of all possibilities to represent any residue modulo N as a sum of M squares (cf. equation (4.5)), hence it equals N^M. And

$$1 + \sum_{r=0}^{N-1} \rho(r)e^{jr\frac{2\pi i}{N}} = 1 + (N^{M-1} - 1) + \sum_{r=1}^{N-1}\left(N^{M-1} + \left(\frac{r}{N}\right)N^{\frac{M-1}{2}}\right)e^{jr\frac{2\pi i}{N}}$$

$$= N^{M-1} + N^{M-1}\sum_{r=1}^{N-1} e^{jr\frac{2\pi i}{N}} + N^{\frac{M-1}{2}}\sum_{r=1}^{N-1}\left(\frac{r}{N}\right)e^{jr\frac{2\pi i}{N}}$$

$$= N^{M-1} + N^{M-1}(-1) + N^{\frac{M-1}{2}}i\sqrt{N}\left(\frac{j}{N}\right) = i\sqrt{N}^M\left(\frac{j}{N}\right).$$

Hence we have shown that for $N = 7$ and $N = 23$

$$\Theta + \sum_{r=0}^{N-1} \rho(r)\Theta_r = (-i\sqrt{N})^M \Theta|_{(S,+,\frac{NM-M}{2})}$$

An analogous calculation shows that this formula is correct for $N = 2, 3, 5, 11$ as well. The proof is a simplification of the argument above as the numbers $\rho(r)$ are equal for all non-zero r.

Furthermore,

$$(-i\sqrt{N})^M \Theta|_{(S,+,\frac{NM-M}{2})} - \Theta = \sum_{r=0}^{N-1} \rho(r)\Theta_r$$

From theorem 3.3 and chapter 4.2 (in particular theorem 4.2) we know that

$$(-i\sqrt{N})^M \theta_L|_{(S,+,\frac{NM-M}{2})} - \theta_L = \theta_{L^*} - \theta_L = \sum_{r=0}^{N-1} \rho(r)\theta_{L+\lambda_r^*}$$

where $\theta_{L+\lambda_r^*}$ is the theta-function of the lattice L translated by an element $\lambda_r^* \in L^*$ such that $\frac{(\lambda_r^*)^2}{2} \equiv \frac{-r}{N} \mod \mathbb{Z}$. Now, if $\lambda \in L$, $\frac{1}{2}(\lambda_r^* + \lambda)^2 = \frac{(\lambda_r^*)^2}{2} + (\lambda_r^*, \lambda) + \frac{\lambda^2}{2} \in \mathbb{Z} - \frac{r}{N}$. This shows that $\theta_{L+\lambda_r^*}$ has a q-expansion with exponents in $\mathbb{Z} - \frac{r}{N}$ only, just as Θ_r. This proves that $\theta_{L+\lambda_r^*} = \Theta_r$ for all r and concludes the proof of theorem 1.7. $\square$

CHAPTER 5

The Real Simple Roots

Let σ be an automorphism of the Leech lattice Λ of order N and cycle shape $1^M N^M$, as before. In this chapter we determine the real simple roots of the GKM $\mathcal{G}_N$ constructed from σ. As the imaginary simple roots of $\mathcal{G}_N$ were explicitly described in theorem 1.6, this will therefore complete the explicit identification of all simple roots of $\mathcal{G}_N$, and enable us to calculate all entries into the generalized Cartan matrix of $\mathcal{G}_N$. We identify the set of real simple roots with a set $\mathcal{R}$ which consists of the fixed point lattice Λ^σ and some elements of its dual $\Lambda^{\sigma*}$. We prove a number of results concerning the decomposition of space induced by the set $\mathcal{R}$. This programme is a generalization of work by Borcherds. In [**Bor92**] the real simple roots of the monster Lie algebra are identified with the 24-dimensional Leech lattice. The holes of the Leech lattice, corresponding to the finite and affine subalgebras of the monster Lie algebra, are enumerated in [**CS88**], chapter 25. Returning to the GKMs $\mathcal{G}_N$, the set $\mathcal{R}$ does not form a lattice though many of the properties are preserved. In particular, the decomposition of space will only produce holes which have finite and affine diagrams. We prove that they provide the complete classification of all finite and affine subalgebras of the GKM $\mathcal{G}_N$. The results of chapter 5 will be used in chapter 6 to carry out the explicit decomposition of space into finite and affine holes. This in turn enables us to identify hyperbolic subalgebras of the GKM $\mathcal{G}_N$ and thus to obtain upper bounds for their root multiplicities.

As identified in theorem 1.6, the real simple roots of the GKM $\mathcal{G}_N$ are the simple roots of its Weyl group. In section 5.1 we determine the set of simple roots of the Lorentzian lattice explicitly. We identify this set with the set $\mathcal{R}$. Section 5.2 will show how subsets of $\mathcal{R}$ translate to Dynkin diagrams and which Dynkin diagrams can be expected. As a by-product, we will be able to give an explicit description of the generalized Cartan matrix of $\mathcal{G}_N$. Section 5.3 establishes a number of results on the volumes of the holes corresponding to those Dynkin diagrams. Section 5.4 looks into the symmetries and automorphisms of the holes of $\mathcal{R}$. These will be used to establish relations between the hole decomposition of the Leech lattice and the hole decompositions of interest in this work.

5.1. The Set of Real Simple Roots

This section identifies the real simple roots of the GKM $\mathcal{G}_N$ associated with an automorphism σ of the Leech lattice Λ of order N, as constructed in the theorem 1.6. We consider all relevant orders $N = 2, 3, 5, 7, 11, 23$. Let, as before, Λ^σ denote the fixed point lattice, $L = \Lambda^\sigma \oplus II_{1,1}$ the corresponding Lorentzian lattice. Let $\pi_{\Lambda^\sigma} : \Lambda \to \Lambda^\sigma$ and $\pi_L : \Lambda \oplus II_{1,1} \to L$ be the respective projections. Formula (1.22) expressed the multiplicities of roots in terms of the trace of σ on the root space E_r (see section 1.5). Here r is an element of W^σ, the Weyl group of the fixed point lattice $\Lambda^\sigma \oplus II_{1,1}$, which was characterised in lemma 1.1.

By lemma 1.3 and 1.2a, the twisted Weyl group W^σ is the full reflection group of L. Lemma 1.2b and 1.2c give an explicit description of the simple roots of L.

THEOREM 5.1. *The (primitive) simple roots of the lattice L, and thus the (primitive) real simple roots of the GKM $\mathcal{G}_N$, are the following:*

(5.1) $$\left(\lambda, 1, \frac{\lambda^2}{2} - 1\right) \quad \text{for} \quad \lambda \in \Lambda^\sigma$$

and

(5.2) $$\left(\lambda, N, \frac{\lambda^2}{2N} - 1\right) \quad \text{for} \quad \lambda \in N\Lambda^{\sigma*} \quad \text{such that} \quad N \mid \left(\frac{\lambda^2}{2N} - 1\right).$$

The former simple roots have height 1 and norm 2 whereas the latter have height N and norm $2N$.

PROOF. We begin by determining all roots of L. As introduced in the context of lemma 1.2, we only consider primitive roots. Suppose, $r \in L$ is a (primitive) root of L. By definition, the norm of r must be greater than 0. Because r is primitive, if $n \in \mathbb{Z}$, $|n| > 1$, then $\frac{r}{n}$ is not in L. We recall that the dual lattice of L satisfies $L^* = (\Lambda^\sigma)^* \oplus II_{1,1}$. For any $v \in L$, the inner product (r, v) is integer, hence the greatest common divisor $d = (r, L)$ of the absolute values of all inner products (r, v), $v \in L$, is defined. It follows that $\frac{r}{d} \in L^*$. Lemma 4.1 implies that $NL^* \subset L$. As N is prime the only possible cases are $d = 1$ or $d = N$. If $d = (r, L) = 1$ then r^2 must necessarily be equal to 1 or 2 because $2(r, v)/(r, r)$ must be integer for all $v \in L$. As L is even, only the case $r^2 = 2$ occurs. Hence, this case yields precisely all $r \in L$ of norm 2 as roots. The remaining case, $d = (r, L) = N$, implies that $r^2 = N$ or $r^2 = 2N$. For $N \neq 2$, the case $r^2 = N$ is impossible because N is odd and L even. We conclude that $r \in NL^*$ will be a primitive root if and only if $r^2 = 2N$. This determines the primitive roots of L.

We now turn to the identification of the simple roots of L, which, at the same time, are the real simple roots of $\mathcal{G}_N$ following theorem 1.6. We use the description of the simple roots of W^σ in lemma 1.2 to derive formulas (5.1) and (5.2). Theorem 1.6 established that the Weyl vector ρ has the coordinates $(0, 0, 1)$. Let $r = (\lambda, h, n)$ be a (primitive) simple root of W^σ, so of norm 2 or norm $2N$ by the first part of the proof. If N does not divide h then $r \notin NL^*$, hence $(r, L) = 1$. As seen in the first part of this proof this implies $r^2 = 2$, and hence lemma 1.2b requires $(r, \rho) = -1$ which is $h = 1$. From $r^2 = 2$ we conclude that r is of the type of formula (5.1). The root r in formula (5.1) furthermore satisfies the scaling condition of lemma 1.2c so that r is the primitive simple root identified in lemma 1.2.

If, on the other hand N does divide h, then $h = Nh'$ and lemma 1.2b implies that $-(\rho, r) = Nh'$ divides (r, v) for all $v \in L$; hence we conclude that $r \in h'NL^*$, and $\frac{r}{h'} \in NL^* \subset L$. Thus r will only be primitive if $h' = 1$. As seen in the first part of this proof, this implies $r^2 = 2N$. Hence r is of the type of formula (5.2). The root r in formula (5.2) furthermore satisfies the scaling condition of lemma 1.2c so that r is the primitive simple root identified in lemma 1.2. □

If $(\lambda, N, N*)$ is a real simple root of norm $2N$ then $(\frac{\lambda}{N}, 1, *)$ is an element of L^* that defines the same reflection. In particular, $\frac{\lambda}{N}$ is an element of $(\Lambda^\sigma)^*$ of norm $\frac{2}{N}$ mod $2\mathbb{Z}$. Conversely, theorem 5.1 shows that any such element of the dual represents a real simple root of the GKM $\mathcal{G}_N$. Let us introduce the notation $\mathcal{R}_{dual}$ for those elements of the dual lattice which represent real simple roots of norm

$2N$. Further, we will write $\mathcal{R}_{fix}$ for the fixed point lattice when it is understood to represent the real simple roots of norm 2 of the GKM $\mathcal{G}_N$. The totality of all real simple roots of the GKM $\mathcal{G}_N$ will then be represented by the set $\mathcal{R}$ which is the disjoint union of $\mathcal{R}_{fix}$ and $\mathcal{R}_{dual}$.

We have the following result of [**Bor90a**] about the geometry of the fixed point lattice:

THEOREM 5.2. *The fixed point lattice Λ^σ can be covered by spheres of radius $\sqrt{2}$ centered at the elements of $\mathcal{R}_{fix}$ plus spheres of radius $\sqrt{\frac{2}{N}}$ centered at the elements of $\mathcal{R}_{dual}$.*

PROOF. This is a direct consequence of theorem 3.1 of [**Bor90a**]. The affine space described there can be identified with the fixed point lattices considered in this work. Then, the radius of the spheres is generally identified as $\sqrt{\left(\frac{r}{(r,\rho)}\right)^2}$. □

REMARK. We can understand this result as follows. We recall from chapter 4 that the shortest non-zero vectors $\lambda^* = \pi_{(\Lambda^\sigma)^\perp}(\lambda)$ of $\Lambda^{\sigma\perp *}$ have norm $2 - \frac{2}{N}$. This implies $\pi_{\Lambda^\sigma}(\lambda)$ is an element of $\mathcal{R}_{dual}$. Hence, if we intersect a covering of the Leech lattice (spheres of radius $\sqrt{2}$) with the span of Λ^σ we obtain precisely the kind of covering as described in theorem 5.2.

We furthermore observe that - even though the set of real simple roots $\mathcal{R}$ is not a lattice - the group of automorphisms of this set is the group of automorphisms of the fixed point lattice $\mathcal{R}_{fix}$.

5.2. Holes and Dynkin Diagrams

Given a GKM with (symmetric) Cartan matrix C and simple roots $\{r_i\}$, we may define a diagonal matrix $D = (d_{ij})$ such that the diagonal elements of the matrix $A = DC$ are either equal to 2, or less or equal zero. We may now associate a Dynkin diagram with any subset of simple roots which consists entirely of real simple roots. We use the conventions for Dynkin diagrams laid out in [**Wan91**], because in chapter 6 we will relate the results of this work to the classifications in [**Wan91**].

We will use the following conventions: The Dynkin diagram of a subalgebra contains one node for every real simple root. Suppose the submatrix of the matrix A, corresponding to the real simple roots r_i and r_j is of form $\begin{pmatrix} 2 & a_{ij} \\ a_{ji} & 2 \end{pmatrix}$ then their respective nodes in the Dynkin diagram will be linked by $\max(|a_{ij}|, |a_{ji}|)$ bonds with additional arrows pointing towards i if $|a_{ij}| > 1$. Note that this notation does contain all information about the Cartan matrix in the cases of the finite, affine, and hyperbolic diagrams we will use it for. In particular, $\begin{pmatrix} 2 & -2 \\ -2 & 2 \end{pmatrix}$ corresponds to a double bond with arrows pointing to both sides. $\begin{pmatrix} 2 & -1 \\ -1 & 2 \end{pmatrix}$ corresponds to a single bond. Note further that the conventions imply that one-sided arrows point toward the shorter root should there be one.

The Dynkin diagram has so far been associated to the real simple roots of the GKM $\mathcal{G}_N$. We will now establish how to associate it directly to subsets of $\mathcal{R} = \mathcal{R}_{fix} \cup \mathcal{R}_{dual}$. There are three cases.

Case A: Both real simple roots are of norm 2. Then $r_1 = \left(\lambda, 1, \frac{\lambda^2}{2} - 1\right)$ and $r_2 = \left(\mu, 1, \frac{\mu^2}{2} - 1\right)$. They correspond to $\lambda, \mu \in \mathcal{R}_{fix}$ and

$$(5.3a) \qquad r_1 \cdot r_2 = \lambda \cdot \mu - \left(\frac{\lambda^2}{2} - 1\right) - \left(\frac{\mu^2}{2} - 1\right) = 2 - \frac{1}{2}(\lambda - \mu)^2$$

Thus the two nodes will be unlinked if $(\lambda - \mu)^2 = 4$, they will be linked by one bond if $(\lambda - \mu)^2 = 6$, they will be linked by a double bond with arrows on both sides if $(\lambda - \mu)^2 = 8$. Other types of bond will not arise in this work.

Case B: Both real simple roots are of norm $2N$. Then $r_1 = \left(\lambda, N, \frac{\lambda^2}{2N} - 1\right)$ and $r_2 = \left(\mu, N, \frac{\mu^2}{2N} - 1\right)$. They correspond to $\frac{\lambda}{N}, \frac{\mu}{N} \in \mathcal{R}_{dual}$ and

$$(5.3b) \quad r_1 \cdot r_2 = \lambda \cdot \mu - \left(\frac{\lambda^2}{2} - N\right) - \left(\frac{\mu^2}{2} - N\right)$$
$$= 2N - \frac{1}{2}(\lambda - \mu)^2 = \frac{N^2}{2}\left[\frac{4}{N} - \left(\frac{\lambda}{N} - \frac{\mu}{N}\right)^2\right].$$

Thus the two nodes will be unlinked if $\left(\frac{\lambda}{N} - \frac{\mu}{N}\right)^2 = \frac{4}{N}$, they will be linked by one bond if $\left(\frac{\lambda}{N} - \frac{\mu}{N}\right)^2 = \frac{6}{N}$, they will be linked by a double bond with arrows on both sides if $\left(\frac{\lambda}{N} - \frac{\mu}{N}\right)^2 = \frac{8}{N}$. Other types of bond will not arise in this work.

Case C: The real simple root r_1 is of norm 2, r_2 is of norm $2N$. Then $r_1 = \left(\lambda, 1, \frac{\lambda^2}{2} - 1\right)$ and $r_2 = \left(\mu, N, \frac{\mu^2}{2N} - 1\right)$. They correspond to $\lambda \in \mathcal{R}_{fix}$, $\frac{\mu}{N} \in \mathcal{R}_{dual}$ and

$$(5.3c) \qquad r_1 \cdot r_2 = \lambda \cdot \mu - N\left(\frac{\lambda^2}{2} - 1\right) - \left(\frac{\mu^2}{2N} - 1\right) = N + 1 - \frac{N}{2}\left(\lambda - \frac{\mu}{N}\right)^2.$$

Thus the two nodes will be unlinked if $\left(\lambda - \frac{\mu}{N}\right)^2 = 2 + \frac{2}{N}$, that is, if they are of minimal distance; they will be linked by N bonds with an arrow pointing toward r_1 if $\left(\lambda - \frac{\mu}{N}\right)^2 = 4 + \frac{2}{N}$. Note that for a fixed element of the dual $\frac{\mu}{N}$ this exactly describes the set of lattice elements closest but one to $\frac{\mu}{N}$. Other types of bond will not arise in this work.

The above calculations complete, as a corollary, the explicit specification of the elements of the generalized Cartan matrices for the GKMs constructed in this work.

THEOREM 5.3. *The generalized Cartan matrix C_N of the GKM $\mathcal{G}_N$ has been completely and explicitly determined.*

PROOF. The set of imaginary simple roots has been identified in theorem 1.6, the set of real simple roots in theorem 5.1. The products of any two real simple roots have been determined in equations (5.3). The remaining products involving imaginary simple roots, that is positive multiples $n\rho, m\rho$ of the Weyl vector ρ are as follows;

$$(n\rho, m\rho) = 0,$$
$$(5.4) \qquad \left(n\rho, \left(\lambda, 1, \frac{\lambda^2}{2-1}\right)\right) = -n,$$
$$\left(n\rho, \left(\lambda, N, \frac{\lambda^2}{2N-1}\right)\right) = -nN.$$

This completes the explicit specification of the generalized Cartan matrix. □

We need to recall some facts about Lie algebras from [**Kac90**]. Given a system of n simple roots α_1, ..., α_n of a Lie algebra with Cartan matrix $C = (a_{ij})$, there is a system of co-roots $\alpha_1^\vee$, ..., $\alpha_n^\vee$ defined by

$$(5.5) \qquad \langle \alpha_i^\vee, \alpha_j \rangle = a_{ij} \quad (i,j = 1, \ldots, n)$$

We define a scalar product in the space of roots $[\alpha]$ and the space of co-roots, $[\alpha^\vee]$. We can then find an isomorphism $\nu : [\alpha^\vee] \to [\alpha]$ such that

$$(5.6) \qquad \nu(\alpha_i^\vee) = \frac{2}{(\alpha_i, \alpha_i)} \alpha_i.$$

We may form a Cartan matrix from a subset of the real simple roots of the GKM $\mathcal{G}_N$. In that situation, the above works out as follows: $\alpha_i^2 = 2$ implies $(\alpha_i^\vee)^2 = 2$, and $\alpha_i^2 = 2N$ implies $(\alpha_i^\vee)^2 = \frac{2}{N}$. We further recall that for any simple finite-dimensional Lie algebra there exists a Weyl vector. This is a vector ρ of the root space satisfying

$$(5.7a) \qquad (\rho, \alpha_i) = \frac{(\alpha_i, \alpha_i)}{2} \quad \text{for all } i.$$

Unlike section 1.1.3, in this case, we use the standard sign convention for the Weyl vector to simplify comparison with the statements of [**Kac90**]. If we define $\rho^\vee := \nu^{-1}\rho$ then

$$(5.7b) \qquad (\rho^\vee, \alpha_i^\vee) = \left(\nu^{-1}\rho, \nu^{-1}\left(\frac{2}{\alpha_i^2}\alpha_i\right) \right) = \frac{2}{\alpha_i^2}(\rho, \alpha_i) = 1.$$

We can express $\rho^\vee$ as an integer linear combination of the simple co-roots: $\rho^\vee = \sum_i n_i^\vee \alpha_i^\vee$. Then

$$(5.8) \qquad \rho^2 = \rho^{\vee 2} = \sum_i n_i^\vee (\rho^\vee, \alpha_i^\vee) = \sum_i n_i^\vee.$$

For any simple Lie algebra of affine type there exists a vector δ such that

$$(5.9a) \qquad \delta^2 = (\delta, \alpha_i) = 0 \quad \text{for all } i.$$

δ can be expressed as a linear combination of the simple roots: $\delta = \sum_i n_i \alpha_i$. δ is then characterised by the condition that the n_i are integer with greatest common divisor 1. Recall from section 1.1.2 that, in the framework of [**Jur96**], we consider roots in a root space which is extended by degree derivations. The vector δ will then have non-zero components in the degree derivations (which are part of the null space of the bilinear form). Similarly, there exists a vector $\delta^\vee = \sum_i n_i^\vee \alpha_i^\vee$ such that

$$(5.9b) \qquad \delta^{\vee 2} = (\delta^\vee, \alpha_i^\vee) = 0 \quad \text{for all } i,$$

uniquely determined by the condition that the $n_i^\vee$ are integer with greatest common divisor 1. If $\delta = \sum n_i \alpha_i$ let d denote the greatest common divisor of the integers $n_i \frac{\alpha_i^2}{2}$. We find that

$$(5.10) \qquad \delta^\vee = \sum_i n_i^\vee \alpha_i^\vee = \frac{1}{d} \nu^{-1} \delta,$$

5.2. HOLES AND DYNKIN DIAGRAMS

We finally recall the definition of the Coxeter number and the dual Coxeter number of an affine Lie algebra:

$$(5.11) \qquad h = \sum_i n_i, \qquad h^\vee = \sum_i n_i^\vee.$$

Next, we need to turn to semisimple finite and affine Lie algebras, that is, direct sums of simple Lie algebras. The roots of different components are mutually orthogonal. Hence, we find that there exist ρ (or δ respectively) as a direct sum of the ρ (or δ) of the components.

We are now in the position to generalize the notion of a hole in a lattice to an analogue in the set $\mathcal{R}$ representing the real simple roots of the GKM $\mathcal{G}_N$. For any point in the $2M$-dimensional space spanned by the fixed point lattice we define a radius function as follows

$$(5.12) \quad \mathrm{r}_\mathcal{R}(x) = \min\{\sqrt{(x-v)^2}, \sqrt{(x-d)^2 + 2 - \frac{2}{N}} \mid v \in \mathcal{R}_{fix},\ d \in \mathcal{R}_{dual}\}.$$

The proof of lemma 5.1 below will clarify the motivation behind this definition for a generalized radius. Because of theorem 5.2, the radius function is bounded above by $\sqrt{2}$. A local maximum of the radius function can then be considered as a generalized hole of the set of real simple roots $\mathcal{R}$. The value of the radius function will be called the generalized radius of the hole. As the dimension of space is $2M$ there will be at least $2M + 1$ real simple roots closest (in the sense of the radius function) to a generalized hole. The real simple roots which are closest in the sense of the radius function, shall be referred to as the generalized vertices of the generalized hole. They will include both elements of $\mathcal{R}_{fix}$ and $\mathcal{R}_{dual}$. They form a Dynkin diagram in the way described above, corresponding to a subalgebra of the GKM $\mathcal{G}_N$.

For the remainder of this work let us introduce some notational conventions with respect to generalized holes. Let H denote the set of generalized vertices of such a hole, let $\langle H \rangle$ denote the convex hull of the vertices, that is the 'hole' in its spatial meaning, and let $\Delta(H)$ denote the Dynkin diagram associated to the hole. We will drop the brackets $\langle \rangle$ occasionally when there is no need to distinguish between a hole and its set of vertices.

LEMMA 5.1. a) *Suppose that we are given a subset H of $\mathcal{R}$ such that $\Delta(H)$ is of finite, or affine, type. Then there exists a (generalized) centre c of (generalized) radius r_0 from all elements of H. c lies within the convex hull $\langle H \rangle$.*

b) *Suppose that H is as in a). Then there exists no element of $\mathcal{R}$ which has smaller (generalized) distance from c than r_0.*

c) *Now suppose that $\Delta(H)$ is of finite type and that H contains n elements where $n \leq 2M$ ($2M$ is the dimension of the fixed point lattice). For any (generalized) radius r, such that $r_0 < r \leq \sqrt{2}$, there exists a $(2M-n)$-dimensional manifold of points which have (generalized) radius r from all elements of H.*

PROOF. a) For any Cartan matrix of finite or affine type we can choose a system of corresponding roots (or co-roots) in Euclidean space. Now in $\mathcal{R}$ we consider vectors of norm 2 and norm $\frac{2}{N}$. Thus we represent these by the relevant co-roots $r_i^\vee = (\lambda_i, 1, *)$, rather than the roots. Let us first consider the affine case. We begin by an explicit identification of the centre c. Let $n_i^\vee, h^\vee$ be defined as in equations (5.10), (5.11). From (5.9) we obtain $\delta^\vee = (\sum_i n_i^\vee \lambda_i, h^\vee, *)$. Let

(5.13a) $$c = \sum_i \frac{n_i^\vee}{h^\vee} \lambda_i$$

Then
$$(c - \lambda_j)^2 = (c - \lambda_j, 0, *)^2 = (\frac{\delta^\vee}{h^\vee} - r_j^\vee)^2 = r_j^{\vee 2}.$$

Hence, c is the generalized centre of a hole of generalized radius $r_0 = \sqrt{2}$. It is well known that the solutions $\delta^\vee$ have coefficients all greater zero which by definition is equivalent to the claim that c lies within the convex hull.

In the case of finite Lie algebras let ρ, $\rho^\vee$ be defined as in equation (5.7). The multiple $\frac{\rho^\vee}{\rho^2}$ is contained within the affine span of the co-roots because of (5.8). Then
$$(\frac{\rho^\vee}{\rho^2} - r_i^\vee)^2 = \frac{\rho^2}{\rho^4} - \frac{2(\rho^\vee, r_i^\vee)}{\rho^2} + (r_i^\vee)^2 = (r_i^\vee)^2 - \frac{1}{\rho^2}.$$

Let us now define

(5.13b) $$c = \frac{\sum_i n_i^\vee \lambda_i}{\rho^2}$$

Then $\frac{\rho^\vee}{\rho^2} = (c, 1, *)$ and thus $(\frac{\rho^\vee}{\rho^2} - r_j^\vee)^2 = (c - \lambda_j, 0, *)^2 = (c - \lambda_j)^2$. Hence we have identified c as the generalized centre in terms of elements of $\mathcal{R}$ and justified the definition of the generalized radius function. This also provides the minimal radius $r_0 = 2 - \frac{1}{\rho^2}$. A case by case analysis shows that the coefficients obtained for the centre again are all greater than zero and thus c is contained in the convex hull $\langle H \rangle$.

b) We begin by considering the following special case: Suppose that the finite diagram H does contain a root of norm 2, say r_1. We will prove that no element p of $\mathcal{R}_{fix}$ lies closer to c than r_0.

Without loss of generality we can assume that the element p is the origin in $\mathcal{R}$ as $\mathcal{R}$ is translation invariant. Thus we consider the following setup: the finite-dimensional Lie algebra is represented by elements of $\mathcal{R}_{fix}$ of norms 4, 6, 8,... and elements of $\mathcal{R}_{dual}$ of norms $2 + \frac{2}{N}$, $4 + \frac{2}{N}$,... We need to show that $c^2 \geq (c - \lambda_1)^2$. This is equivalent to $(2c - \lambda_1, \lambda_1) \geq 0$. We return to the equation $\rho^\vee = \sum n_i^\vee r_i^\vee$ and recall that $r_i^\vee = (\lambda_i, 1, x_i)$. If $(r_i^\vee)^2 = 2$, then $x_i = \frac{\lambda_i^2}{2} - 1 \geq 1$. If $(r_i^\vee)^2 = \frac{2}{N}$, then $x_i = \frac{\lambda_i^2}{2} - \frac{1}{N} \geq \frac{2 + \frac{2}{N}}{2} - \frac{1}{N} = 1$. Now $(\rho^\vee, r_1^\vee) = 1 = \frac{(r_1^\vee, r_1^\vee)}{2}$. This implies

$$(2\frac{\rho^\vee}{\rho^2} - r_1^\vee, r_1^\vee) = 2(\frac{1}{\rho^2} - 1).$$

Rewriting this in terms of the elements λ_i of $\mathcal{R}$ we obtain:

$$\frac{2}{\rho^2} - 2 = \left((2c - \lambda_1, 1, 2\frac{\sum_i n_i^\vee x_i}{\sum_i n_i^\vee} - (\frac{\lambda_1^2}{2} - 1)), (\lambda_1, 1, \frac{\lambda_1^2}{2} - 1) \right)$$

$$= (2c - \lambda_1, \lambda_1) - 2\frac{\sum_i n_i^\vee x_i}{\sum_i n_i^\vee}.$$

$$(2c - \lambda_1, \lambda_1) = \frac{2}{\rho^2} - 2 + 2\frac{\sum_i n_i^\vee x_i}{\sum_i n_i^\vee} \geq \frac{2}{\rho^2} > 0,$$

as every individual x_i was greater or equal 1. This concludes the argument for the special case. There remain the following cases for the finite types: 1) $p \in \mathcal{R}_{fix}$ but all roots of the finite diagram have norm $2N$. 2) $p \in \mathcal{R}_{dual}$. It suffices to remark for these cases that the proof works along exactly the same lines of argument. We only have to adjust the norms of the vectors involved. For the affine cases we use the vector $\delta^\vee$ of equation (5.9) instead of $\rho^\vee$.

c) We now identify the manifold of centres for radius greater than the minimal radius r_0. Suppose that $H = H_f \cup H_d$ with $H_f \subset \mathcal{R}_{fix}$ and $H_d \subset \mathcal{R}_{dual}$. Suppose further that $|H_f| = n_f$ and $|H_d| = n_d$. Because the n_f vectors are affinely independent there exists an affine centre c_f for them. We consider the space $F^\perp$ orthogonal to the affine span of the n_f vectors, passing through c_f. $F^\perp$ has $2M - n_f + 1$ dimensions. Equally, because the n_d dual vectors are affinely independent there exists an affine centre c_d for them. We consider the space $D^\perp$ orthogonal to the affine span of the dual vectors, passing through c_d. $D^\perp$ has $2M - n_d + 1$ dimensions.

The set H was assumed to be affinely independent. Hence the affine spaces $F^\perp$ and $D^\perp$ intersect in a $(2M - n + 2)$-dimensional affine space. Every point of this space has a certain radius r_f which is the distance to any element of H_f, and a certain radius r_d which is the distance to any element of H_d. In turn, all points of this space of a fixed r_f form a sphere in the affine space centered at $\tilde{c}_f$, which is the projection of c_f. Note that there is a minimal $r_{0f} < \sqrt{2}$ which corresponds to the distance of $\tilde{c}_f$ from the elements of H_f. Similarly, the points of the intersection space of distance r_d from the elements of H_d form spheres centered at some $\tilde{c}_d$. Again, there exists a minimal $r_{0d} < \sqrt{\frac{2}{N}}$. Clearly, $\tilde{c}_d \neq \tilde{c}_f$.

An element of the $2M$-dimensional space spanned by the fixed point lattice will have distance r from all elements of H_f and distance $\sqrt{r^2 - 2 + \frac{2}{N}}$ from all elements of H_d if and only if it lies in the intersection of the corresponding spheres

$$S\left(\tilde{c}_f, \sqrt{r^2 - r_{0f}^2}\right) \cap S\left(\tilde{c}_d, \sqrt{r^2 - 2 + \frac{2}{N} - r_{0d}^2}\right).$$

Two spheres in space of different radius and centre can intersect in the following ways:

α) the volumes enclosed by the spheres are disjoint.
β) the volumes enclosed by the spheres intersect in one point.
γ) the intersection is a proper subset of both volumes enclosed, but does not consist of a single point. It then is a $(2M - n)$-dimensional manifold.
δ) the volumes enclosed lie within one another.

Clearly, for the centre c and radius r_0, situation β) holds. For radius $\sqrt{2}$ situation γ) must hold because for instance the elements of the dual lattice and those of the lattice itself have distance greater than 2. Hence we cannot already be in situation δ). From this it follows that situation γ) is attained for any radius between r_0 and $\sqrt{2}$. Thus the intersection provides the manifold as claimed for any radius between the two bounds given in the claim. $\square$

The question whether a hole has covering radius less or equal to $\sqrt{2}$ is significant because of

THEOREM 5.4. *The Dynkin diagram of a generalized hole is of finite type if and only if the hole has generalized radius less than $\sqrt{2}$. The Dynkin diagram is of affine type if and only if the hole has generalized radius equal to $\sqrt{2}$.*

PROOF. The argument is a straightforward generalization of the arguments used in chapter 23 of [**CS88**]. It is well known (see e.g. [**Kac90**], chapter 4) that an indecomposable Cartan matrix is of finite type if and only if all its principal minors are positive definite. Also, an indecomposable Cartan matrix is of affine type if and only if it has determinant 0 and all its proper principal minors are positive definite. We can rephrase this as follows: For any Cartan matrix of finite or affine type, and only for these, we can choose a co-root system corresponding to the Cartan matrix, in Euclidean space. Then the Cartan matrix is of finite type if and only if the co-roots are linearly independent. Then the origin will not lie within the affine span of the simple roots. The origin is, however a point at (generalized) radius $\sqrt{2}$ from all simple roots. Thus the (generalized) centre of the affine span has radius less than $\sqrt{2}$. This proves the claim for the finite case. The argument for the affine case is similar, using the existence of the vector $\delta^\vee$ of equation (5.9). □

We observe, as a corollary of theorem 5.4, that the hole diagrams only contain bonds as described in the first part of section 5.2. In particular, for $N \geq 5$, the elements of $\mathcal{R}_{fix}$ and those of $\mathcal{R}_{dual}$ will form disjoint components of the diagram of a hole because, if $\lambda \in \Lambda^\sigma$ and $\frac{\mu}{N}$ in the dual are both vertices of a generalized hole, then

$$\left(\lambda - \frac{\mu}{N}\right)^2 \leq \left(\sqrt{2} + \sqrt{\frac{2}{N}}\right)^2 = 2 + \frac{2}{N} + \frac{4}{\sqrt{N}} < 4 + \frac{2}{N}.$$

We now use lemma 5.1 to prove that the set of finite and affine subalgebras identified by theorem 5.4 is in fact the complete classification of all finite and affine subalgebras contained in the GKM $\mathcal{G}_N$. Thus the task of identifying subalgebras will be reduced to a classification of generalized holes.

THEOREM 5.5. *Suppose there is given a subset H of $\mathcal{R}$ such that the Dynkin diagram $\Delta(H)$ is of either finite or affine type. Then there exists a generalized hole K in $\mathcal{R}$ such that $H \subset K$.*

PROOF. Given a set H of finite, or affine, type, then by lemma 5.1 we identify its (generalized) centre c and radius r_0. We prove the theorem by induction on the dimension of the affine span of H. Suppose the radius r_0 of H equals $\sqrt{2}$. Then there are no points in the set $\mathcal{R}$ closer to c by lemma 5.1b. Let H' be the collection of all points at (generalized) distance $\sqrt{2}$ from c. Then the affine span of H' must be the total space because $\sqrt{2}$ is the generalized covering radius of the set $\mathcal{R}$. Thus c is the centre of a hole and we have detected the diagram $\Delta(H)$ as part of its diagram.

Now suppose H has radius $r_0 < \sqrt{2}$. If the dimension of the affine span of H is less than $2M$ we consider the manifold $\mathcal{M}(r)$ of points at (generalized) distance $r \geq r_0$ from each point of H. Again, by lemma 5.1b there are no elements of $\mathcal{R}$ closer to c than distance r_0. Hence there will be a minimal $r_1 \geq r_0$ such that for some $c_1 \in \mathcal{M}(r_1)$ there is a further point of $\mathcal{R}$ at that same distance from c_1. (That such an r_1 exists follows from the fact that there exists a generalized covering radius to the lattice. Hence r_1 is bounded from above by $\sqrt{2}$.) We then collect all points at that generalized distance r_1 from c_1 to form the set H_1. Now either $r_1 = \sqrt{2}$.

In this case we are done (see above). Or, $r_1 < \sqrt{2}$. This implies that the extended set of simple roots still is of finite type. If the dimension of the affine span of H_1 equals $2M$, we are done. Else, the assumptions of lemma 5.1 are satisfied. Thus the theorem follows by induction. □

We conclude this section with the observation that any affine component determines the generalized centre of a hole and hence the complete hole. On the other hand, finite components can be part of several holes.

5.3. The Volume Formula

This section establishes that the generalized holes partition the vector spaces spanned by the real simple roots $\mathcal{R}$ of the GKM $\mathcal{G}_N$ and calculates the volumes of all finite and affine holes which will occur in the decomposition. In chapter 6 we will then carry out the explicit decomposition of space for all relevant N. We will have a check of the decomposition because the sum of the volumes of the holes found in a fundamental domain must equal the total volume of this fundamental domain.

We begin by establishing that the holes as described in sections 5.1 and 5.2 actually partition space in the following sense.

LEMMA 5.2. *We consider generalized holes as defined in sections 5.1 and 5.2. If two such holes intersect they do so in a common face, the convex hull of at most $2M$ points, where $2M$ is the dimension of space. Every point of space lies either within such a face or within the interior of a unique (generalized) hole.*

PROOF. The first claim is straightforward. For the second, it remains to show the following: Suppose we are given a generalized hole H and a $(2M-1)$-dimensional face F of it. Claim: If we pass through the interior of F we will enter the interior of another generalized hole.

By lemma 5.1 we know that F has a generalized centre c, generalized radius r_0, and that there exists a 1-dimensional manifold $\mathcal{M}$ of points which have equal generalized distance to all points spanning F. For the points $x \in \mathcal{M}$, let $r_F(x)$ be the generalized distance x to the points spanning F, and let $r_{\mathcal{R}-F}(x)$ be the minimal generalized distance of x to any point of $\mathcal{R} - F$. Here we always use the minimal distance function as defined in formula (5.12) of section 5.2. Then, by lemma 5.1, the set
$$\mathcal{M}(r) = \{x \in \mathcal{M} \mid r_F(x) = r\}$$
is non empty for $r_0 \leq r \leq \sqrt{2}$. The manifold $\mathcal{M}$ is symmetric with respect to the hyperplane spanned by F. Hence the set $\mathcal{M}'(r) = \mathcal{M}(r) - \langle H \rangle$ is non-empty for $r_0 < r \leq \sqrt{2}$. From lemma 5.1b we know that $r_F(x) < r_{\mathcal{R}-F}(x)$ for $x \in \mathcal{M}(r_0)$. That means that no point of $\mathcal{R}$ is closer or equally close to the generalized centre of F than those spanning F. On the other hand $r_{\mathcal{R}}(x) \leq \sqrt{2}$ for all x. Hence there certainly will be points $p \in \mathcal{R}$, $p \notin F$ closer or equally close to any $x \in \mathcal{M}(\sqrt{2})$. Hence there will be a minimal r_1, $r_0 < r_1 \leq \sqrt{2}$ such that there exists a point $x \in \mathcal{M}'(r_1)$ with $r_{\mathcal{R}-F}(x) = r_F(x) = r_1$. That implies that x is the centre of a generalized hole of generalized radius r_1 and concludes the proof. □

We consider the fixed point lattice Λ^σ among the roots $\mathcal{R}$. We define a fundamental region of Λ^σ as in, e.g., [**CS88**]. Let $\text{Aut}(\Lambda^\sigma)$ denote the group of automorphisms of the fixed point lattice fixing the origin. Let $\text{Aut}(H)$ denote the group of automorphisms fixing a generalized hole H. Note that, for every H, $\mathrm{Aut}(H)$ can be identified with a subgroup of $\mathrm{Aut}(\Lambda^\sigma)$ using a unique translation.

As a corollary to lemma 5.2 we obtain a volume formula for the fundamental volume V.

THEOREM 5.6.
$$V = \sum_{\text{holes } H} \mathrm{vol}(H) = |\mathrm{Aut}(\Lambda^\sigma)| \sum_{\text{orbits of holes}} \frac{\mathrm{vol}(H)}{|\mathrm{Aut}(H)|}$$

PROOF. Consider all the holes H within a fundamental region which are equivalent under $\mathrm{Aut}(\Lambda^\sigma)$. Their number is the order of $\mathrm{Aut}(\Lambda^\sigma)$ divided by the number of automorphisms that map H to itself. Lemma 5.2 shows that the total volume V will be accounted for. □

The theory which we developed so far enables us to derive a result concerning the covering radii of the fixed point lattices which provides an interesting insight into the relations between the fixed point lattices Λ^σ and the Leech lattice, even though we will not use it in the remainder of this work.

THEOREM 5.7. *The covering radius of Λ^σ is $\sqrt{2 + \frac{2}{N}}$.*

PROOF. We begin with the cases $N = 2$ and $N = 3$. Here it is well known that the covering radius is as claimed, see, e.g., [**CS88**], chapter 4.

We can now restrict ourselves to the remaining cases $N = 5, 7, 11, 23$. It is straightforward to see that the covering radius of the fixed point lattice must be greater or equal to $\sqrt{2 + \frac{2}{N}}$ because the elements of $\mathcal{R}_{dual}$ have distance greater or equal $\sqrt{2 + \frac{2}{N}}$ from every element of Λ^σ. (See, e.g. equations (5.3).) It remains to show that space is covered by spheres of radius $\sqrt{2 + \frac{2}{N}}$ centered at the elements of Λ^σ. We have established in theorem 5.6 that space is decomposed into convex holes which correspond to finite or affine subalgebras of $\mathcal{G}_N$. Hence it suffices to prove that all holes, which occur in the decomposition, are covered by the above spheres.

Let H be the set of vertices of a such a hole, of finite or affine type. Let $H_f = H \cap \mathcal{R}_{fix}$, and $H_d = H \cap \mathcal{R}_{dual}$. As $\Delta(H)$ is of finite or affine type we can represent the elements of H_f by vectors v_i of norm 2, and the elements of H_d by vectors w_j of norm $\frac{2}{N}$, in Euclidean space. Now in the cases $N = 5, 7, 11, 23$ there exist no finite or affine Lie algebras with bonds between the long and the short roots. This implies that all the v_i are orthogonal to all the w_j.

We now consider the convex hull C of the points $0, v_i, w_j$. (If the original $\Delta(H)$ was of finite type this is a simplex, one face of which is the original hole $\langle H \rangle$. If $\Delta(H)$ was of affine type this is the original hole $\langle H \rangle$.) Let the span of the v_i be denoted V. We show that C is covered by spheres $B(v_i, \sqrt{2 + \frac{2}{N}})$, centered at the v_i. We begin by observing that every w_j is contained in every sphere $B(v_i, \sqrt{2 + \frac{2}{N}})$. Because C is convex, no point of C has a distance from V which is greater than $\sqrt{\frac{2}{N}}$, which is the distance from V of the vertices w_j. Because of the mutual orthogonality of the v_i and the w_j it follows that the projection into V takes every point of C into $C \cap V$. The spheres $B(v_i, \sqrt{2 + \frac{2}{N}})$ contain multi-dimensional cylinders as follows:

a sphere of radius $\sqrt{2}$ in the intersection with V times a sphere of radius $\sqrt{\frac{2}{N}}$ in the orthogonal dimensions. It follows that C will certainly be covered by the spheres $B(v_i, \sqrt{2 + \frac{2}{N}})$ if $C \cap V$ is covered by spheres $B(v_i, \sqrt{2})$. This however is true because the v_i represent a Lie algebra of finite or affine type. □

REMARK. The same argument can also be used to identify the covering radius in the cases $N = 2$ and $N = 3$. The only additional requirement would be to check the diagrams of type b, c, f (finite and affine) case by case as to whether the convex hull of its vertices is covered by spheres centered at the roots of norm 2. Note that by the above theorem we have related the covering radius of the Leech lattice Λ to the radii of the fixed point lattices Λ^σ. Note furthermore that we have established that the deep holes of Λ^σ are precisely the elements of $\mathcal{R}_{dual}$.

The remainder of this section identifies the volumes of the various finite and affine holes that form the partition of space. Some, though not all of the formulas have been documented in [**CS88**], chapter 25. None of the results which will follow in the remainder of this section 5.3 are original, however I do not know of any reference which states or proves them.

We calculate the volumes of the convex hulls of the finite and affine Dynkin diagrams obtained in the decomposition of the root-space $\mathcal{R}$ of the GKM $\mathcal{G}_N$. We consider a finite Dynkin diagram, Δ. The construction of the set $\mathcal{R}$ implies that Δ can be realized by co-roots $r_i^\vee$ of lengths 2 and $\frac{2}{N}$ with the origin not contained in the convex hull of the vectors $r_i^\vee$. We want to determine the $(n-1)$-dimensional volume $v(\Delta)$ of the convex hull of the roots $r_i^\vee$. The volume V of the n-simplex of the points $0, r_1^\vee, \ldots, r_n^\vee$ can be calculated in two ways: First,

$$V = \text{vol}(0, r_1^\vee, \ldots, r_n^\vee) = \frac{1}{n!} \det(r_1^\vee, \ldots, r_n^\vee)$$

Now, for any matrix X, $(\det X)(\det X^T) = \det(XX^T)$. If X consists of columns x_j then the entries of XX^T will be the inner products (x_i, x_j). In the case at hand, this is the Cartan matrix of the Lie algebra Δ symmetrized such that the diagonal elements are either 2 or $\frac{2}{N}$. Let this matrix be denoted DC. Here, D is a diagonal matrix and C is the Cartan matrix of Δ. We conclude that $V = \frac{1}{n!}\sqrt{\det(DC)}$.

Second, V can be calculated as the product of the $(n-1)$-dimensional volume $v(\Delta)$ and the height of the origin above the affine hyperplane spanned by the $r_i^\vee$, divided by the dimension n. The height vector h stands orthogonal on the hyperplane, that is $(h, h - r_i^\vee) = 0$ for all i. It follows that the direction of h is characterised by the fact that $(h, r_i^\vee)$ is constant for all i. We recall that there exists a unique such direction, given by the Weyl vector $\rho^\vee$ of equation (5.7). Hence h is a multiple of $\rho^\vee$. We further know that h is part of the affine hyperplane of the r_i. Using equation (5.8) we conclude that $h = \frac{\rho^\vee}{\rho^2}$. It follows that $V = \frac{1}{n}|h|v(\Delta) = \frac{1}{n}\frac{v(\Delta)}{\sqrt{\rho^2}}$. Hence

$$(5.14) \qquad v(\Delta) = n\sqrt{\rho^2}\, V = \frac{1}{(n-1)!}\sqrt{\rho^2}\sqrt{\det(DC)}.$$

From this we can furthermore derive the following rules for the volumes of diagrams made up from disjoint components Δ_1 and Δ_2. Disjoint components imply that the affine spans are orthogonal. Hence the norms of the Weyl vectors are additive and the determinants are multiplicative.

We now consider an indecomposable affine diagram, Δ_n. Again, it is represented by co-roots $r_i^\vee$ of norm 2 and $\frac{2}{N}$. We consider the vector $\delta^\vee = \sum_i n_i^\vee r_i^\vee$ as constructed in equation (5.9). We label the co-roots $r_i^\vee$, $i = 0, \ldots, n$ such that $n_0^\vee = 1$. (See [**Kac90**], chapter 4 for the fact that this is possible for all affine subalgebras.)

The convex hull of Δ is a simplex as there are $n+1$ points spanning an n-dimensional space. We understand the $r_i^\vee$ to be vectors written as columns. Then the volume can be calculated as

(5.15)
$$\begin{aligned} n!v(\Delta) &= \det(r_1^\vee - r_0^\vee, r_2^\vee - r_0^\vee, \ldots, r_n^\vee - r_0^\vee) \\ &= \det \begin{pmatrix} 1 & 0 & 0 & \ldots & 0 \\ 0 & r_1^\vee - r_0^\vee & r_2^\vee - r_0^\vee & \ldots & r_n^\vee - r_0^\vee \end{pmatrix} \\ &= \det \begin{pmatrix} 1 & 0 & 0 & \ldots & 0 \\ r_0^\vee & r_1^\vee - r_0^\vee & r_2^\vee - r_0^\vee & \ldots & r_n^\vee - r_0^\vee \end{pmatrix} \\ &= \det \begin{pmatrix} 1 & 1 & 1 & \ldots & 1 \\ r_0^\vee & r_1^\vee & r_2^\vee & \ldots & r_n^\vee \end{pmatrix} = \det \begin{pmatrix} h^\vee & 1 & 1 & \ldots & 1 \\ \delta^\vee & r_1^\vee & r_2^\vee & \ldots & r_n^\vee \end{pmatrix} \\ &= h^\vee \det(r_1^\vee, r_2^\vee, \ldots, r_n^\vee) = h^\vee \sqrt{\det(DC)} \end{aligned}$$

Here, the last step is precisely as in the case of finite diagrams, because the co-roots $r_1^\vee, \ldots, r_n^\vee$ correspond to a finite diagram. Note that the choice of the co-root labelled 0 is not always unique, the only condition we used was that $n_0^\vee = 1$ in the construction of $\delta^\vee$. Still, the result is the same for any such co-root. For disconnected Dynkin diagrams we observe that orthogonality implies that the volumes are multiplicative: $(n_1 + n_2)!v(\Delta_1 \Delta_2) = n_1!v(\Delta_1)n_2!v(\Delta_2)$.

We note that the simply laced diagrams do appear within the decomposition in two varieties. Let Δ_n denote an indecomposable simply laced diagram of rank n such that all roots have norm 2. Here, Δ_n may be either finite or affine. Then $N\Delta_n$ shall denote the diagram of type Δ_n such that all roots have norm $2N$. For the symmetrized Cartan matrices of the co-roots we obtain: $DC(N\Delta_n) = \frac{1}{N}DC(\Delta_n)$. This proves the following formulas for the determinant and the norm of the Weyl vector:

(5.16)
$$\det(N\Delta_n) = \frac{1}{N^n}\det(\Delta_n)$$

(5.17)
$$\rho^2(N\Delta_n) = N\rho^2(\Delta_n).$$

This concludes the results needed to calculate the volumes of the generalized holes. The actual values of determinant and norm of the Weyl vector are well documented in the literature. For convenience, we reproduce in table 5.1 the values for those finite and affine Lie algebras that will occur in chapter 6.

We conclude this section with a remark which will provide a further useful check for forthcoming calculations. If all the co-ordinates of the vertices of a hole are rational so will be the volume. Thus the terms under the square root will have to prove to be squares. Similarly, if M co-ordinates are rational, and M co-ordinates are elements of $\sqrt{N}\mathbb{Q}$ then the volume will be element of $\sqrt{N}^M\mathbb{Q}$.

5.4. THE AUTOMORPHISM GROUPS

diagram	ρ^2	det	diagram	$h^\vee$	det
a_n	$\frac{1}{12}n(n+1)(n+2)$	$n+1$	A_n	$n+1$	$n+1$
b_n	$\frac{1}{6}n(2n-1)(2n+1)$	2^{2-n}	B_n	$2n-1$	2^{2-n}
c_n	$\frac{1}{6}n(n+1)(2n+1)$	1	C_n	$n+1$	1
d_n	$\frac{1}{6}(n-1)n(2n-1)$	4	D_n	$2n-2$	4
e_6	78	3	E_6	12	3
e_7	$\frac{399}{2}$	2	E_7	18	2
e_8	620	1	E_8	30	1
f_4	78	$\frac{1}{4}$	F_4	9	$\frac{1}{4}$
g_2	14	$\frac{1}{3}$	G_2	4	$\frac{1}{3}$
			$A^{(2)}_{2n-1}$	$2n$	1
			$D^{(2)}_{n+1}$	$2n$	2^{2-n}
			$E^{(2)}_6$	12	$\frac{1}{4}$
			$D^{(3)}_4$	6	$\frac{1}{3}$

TABLE 5.1. Parameters of finite and affine Lie algebras

5.4. The Automorphism Groups

The aim of chapter 6 will be to provide a complete account of the types of generalized holes contained in the root system $\mathcal{R}$. Two holes will only be accepted as equivalent if there exists an automorphism of the fixed point lattice taking one to the other. Now it turns out that in higher dimensions the Dynkin diagram formed by the vertices of a hole does not necessarily determine the equivalence class of the hole uniquely. There are examples of such holes in the Leech lattice (see [**CS88**], chapter 25). As we will see in chapter 6, the set $\mathcal{R}$ for $N = 3$ does provide further examples. Hence, there is no hope to prove a general result which lifts holes in the set $\mathcal{R}$ to holes in Λ accounting for equivalence. Therefore, this section will describe a number of results and techniques which will enable us to assert the equivalence of holes on a case by case basis in chapter 6.

We will need the following description of the faces of generalized holes:

LEMMA 5.3. *Let H be a subset of $\mathcal{R}$ such that $\Delta(H)$ is of finite type. Then $\langle H \rangle$ has $2M + 1$ faces, obtained by removing any one of its vertices. Let H be a subset of $\mathcal{R}$ such that $\Delta(H)$ is of affine type. Any face of $\langle H \rangle$ is the convex hull of a subset of its vertices, chosen to the following rule: Remove one vertex (simple root) of every component of the Dynkin diagram.*

REMARK. Note that the vertices of a face of a hole of affine type form a Dynkin diagram of finite type.

PROOF. For the finite type, we remember that the generalized hole is a simplex. For the affine type we observe that we must delete at least one vertex of every component because otherwise the centre of the whole generalized hole would be contained in the span. On the other hand, that leaves at most $2M$ vertices. They will all be needed to form the face in a $2M$-dimensional space. □

The purpose of this section is to identify the relations between holes in the Leech lattice (as listed in [**CS88**], chapter 25) and generalized holes in the sets $\mathcal{R}$. We consider the set of vertices H of a generalized hole $\langle H \rangle$ in the decomposition of $\mathcal{R}$. As before, write H as the disjoint union $H_f \cup H_d$ with $H_f \subset \mathcal{R}_{fix}$ and $H_d \subset \mathcal{R}_{dual}$. Now consider $\lambda^* \in H_d$. Theorem 3.1 shows that the elements of the dual lattice are images under the projection π_{Λ^σ} onto the space spanned by the fixed point lattice. We consider the set of preimages $\pi^-_{\Lambda^\sigma}(\lambda^*)$. If $\lambda \in \Lambda$ such that $\pi_{\Lambda^\sigma}\lambda = \lambda^*$ then $(\lambda - \pi_{\Lambda^\sigma}\lambda)^2 \geq 2 - \frac{2}{N}$ because $(\lambda^*)^2 \equiv \frac{2}{N}$ mod $2\mathbb{Z}$. We consider a dual vector of type $\lambda^* = \frac{1}{\sqrt{8}}(4, 0^{(M-1)}, \frac{4}{N}^{(N)}, 0^{(N)}, ..., 0^{(N)})$ in the notation of chapter 4.1. It has norm $2 + \frac{2}{N}$. Any $\lambda \in \Lambda$ such that

$$(5.18) \qquad \pi_{\Lambda^\sigma}(\lambda) = \lambda^* \text{ and } (\lambda - \pi_{\Lambda^\sigma}\lambda)^2 = 2 - \frac{2}{N}$$

thus has norm 4. Obviously there are precisely N such, namely

$$(5.19) \qquad \lambda_n = (4, 0^{(M-1)}, 0^{n-1}, 4, 0^{(N-n)}, 0^{(N)}, \ldots, 0^{(N)}),$$

for $n = 1, \ldots, N$. For a general dual vector λ^* we recall from chapter 4.4 (see proof of theorem 4.2) that any two vectors of norm $\frac{2}{N}$ modulo $2\mathbb{Z}$ are equivalent modulo Λ^σ. Hence the preimages at distance $2 - \frac{2}{N}$ of any element of $\mathcal{R}_{dual}$ form a cycle of N points, and $\sum_{n=0}^{N-1} \frac{1}{N}\sigma^n(\lambda) = \lambda^*$. Note that this does not imply that all roots of norm $2 + \frac{2}{N}$ are equivalent under the automorphism group of Λ^σ. This is, in fact, not true in the cases $N = 11, 7, 5$.

For the generalized hole $\langle H \rangle$ with set of vertices H and generalized centre c we now consider the following set:

$$(5.20) \qquad H_\Lambda = \{\lambda \in \Lambda \mid \lambda \in H \text{ or } \lambda \text{ satisfies } (5.18) \text{ for } \lambda^* \in H \}$$

From the construction of c in formula (5.13) it is immediate that c has distance less or equal to 2 from all points of H_Λ. Theorem 5.4 holds for the Leech lattice (see [**CS88**], chapter 25). Hence the points of H_Λ are part of either a finite, or an affine, diagram of the Leech lattice.

Suppose that $\Delta(H)$ was of finite type in Λ^σ such that $|H_f| = n$ and (consequently) $|H_d| = 2M+1-n$. First, assume that $n \leq M$. The set H_Λ then will contain $n + N \times (2M+1-n) = MN + n + MN - nN + N = 24 - M + n + (M-n)N + N = 24 + (N-1)(M-n) + N \geq 24 + N$ elements. Hence we have found a hole of Λ of more than 25 vertices and a radius strictly smaller than $\sqrt{2}$. This contradicts the fact that the roots of finite diagrams are affinely independent. Next, assume that $n = M + 1$. Then H_Λ contains $M + 1 + NM = M(N+1) + 1 = 25$ vertices. Thus this describes a complete finite hole H_Λ of the Leech lattice which thus is unique. Further, it follows that σ must preserve the hole H_Λ, or equivalently that $\sigma \in \text{Aut}(H_\Lambda)$.

Now suppose that $\Delta(H)$ is of finite type, and that $n > M + 1$. Then H_Λ has less than 25 elements and hence forms part of faces of a number of finite and affine holes. There is no uniqueness, and σ acts by permuting cycles of these holes which share the fixed vertices.

Finally, suppose that $\Delta(H)$ was of affine type in Λ^σ. Then the distance of the generalized centre c was exactly $\sqrt{2}$ from all points of H_Λ. It then follows from the analogue of theorem 5.4 that there exists an affine hole H_Λ of the Leech lattice with the same centre c. This hole is unique in so far as c lies within its interior. This implies that σ must preserve the hole H_Λ, or equivalently that $\sigma \in \operatorname{Aut}(H_\Lambda)$. This, in turn, implies that all vertices in H_Λ will be preimages of vertices of H. This proves

THEOREM 5.8. *The generalized holes of $\mathcal{R}$ are of the following three types:*
(I) *The interior of each affine hole of $\mathcal{R}$ lies within the interior of a unique affine hole H_Λ of the Leech lattice such that $\sigma \in \operatorname{Aut}(H_\Lambda)$.*
(II) *The interior of each finite hole with $M+1$ elements of $\mathcal{R}_{fix}$ among its vertices lies within the interior of a unique finite hole H_Λ of the Leech lattice such that $\sigma \in \operatorname{Aut}(H_\Lambda)$.*
(III) *Finite holes with more than $M+1$ elements of $\mathcal{R}_{fix}$ among their vertices lie within the faces of some affine and finite holes of the Leech lattice.* □

REMARK. The faces of affine holes correspond to finite diagrams, hence there are no 'affine faces'. That there exist holes of each of the three types will become apparent in the explicit decomposition, as provided in appendix A.

It is also worth noting that the preimages of an element of $\mathcal{R}_{dual}$ form a diagram of type a_1^N, as can be seen in the explicit case considered above (equation (5.19)). Now consider a hole H_Λ of the Leech lattice such that $\sigma \in \operatorname{Aut}(H_\Lambda)$. The action of σ is to permute a number of vectors which belong to the same cycle, and to fix the rest. Thus σ induces a map taking $\Delta(H_\Lambda)$ to the Dynkin diagram of some generalized hole of $\mathcal{R}$. We want to describe this action directly as an action on the diagrams. It is obvious that σ identifies some sets of nodes, and fixes the rest. We need to understand the action on the bonds. First, consider the case when $\lambda \in \Lambda$ is fixed, and μ_i, $i = 1, \ldots, N$ are cyclically permuted by σ. It follows that all $(\lambda - \mu_i)^2$, $i = 1, \ldots, N$ must be equal. From the construction it follows immediately that the vectors λ, μ_j, and $\pi(\mu_j) = \frac{1}{N}\sum_i \mu_i$ form an rectangular triangle. Hence, $(\lambda - \frac{1}{N}\sum_i \mu_i)^2 = (\lambda - \mu_j)^2 - (2 - \frac{2}{N})$. Thus, if λ and μ_j are unlinked in Λ, so are λ and $\frac{1}{N}\sum_i \mu_i$ in $\mathcal{R}$. And, if λ and μ_j are linked in Λ by a single bond, then λ and $\frac{1}{N}\sum_i \mu_i$ will be linked by an arrow with N bonds. Now consider the case of two cycles. Suppose, λ_i, $i = 1, \ldots, N$, and μ_i, $i = 1, \ldots, N$ are such. Name the corresponding simple roots of the monster Lie algebra $r_i = (\lambda_i, 1, \frac{\lambda_i^2}{2} - 1)$ and $s_i = (\mu_i, 1, \frac{\mu_i^2}{2} - 1)$. As observed above, each cycle forms an a_1^N diagram. Hence $(\lambda_i - \lambda_j)^2 = 4$ for all $i, j = 1, \ldots, N$. Hence,

$$(\lambda_i, \lambda_j) = \frac{\lambda_i^2 + \lambda_j^2 - (\lambda_i - \lambda_j)^2}{2} = \frac{\lambda_i^2 + \lambda_j^2}{2} - 2$$

$$(\sum_i \lambda_i)^2 = \sum_{i,j}(\lambda_i, \lambda_j) = \sum_i \sum_{j \neq i}(\frac{\lambda_i^2 + \lambda_j^2}{2} - 2) + \sum_i \lambda_i^2 = N\sum_i \lambda_i^2 - 2N(N-1).$$

Hence, we obtain

$$\frac{(\sum_i \lambda_i)^2}{2N} - 1 = \sum_i \frac{\lambda_i^2}{2} - N.$$

We recall that the root r of norm $2N$ had the form $r = (\sum_i \lambda_i, N, \frac{(\sum_i \lambda_i)^2}{2N} - 1)$. Thus we can write

$$r = (\sum_i \lambda_i, N, \sum_i \frac{\lambda_i^2}{2} - N) = \sum_i (\lambda_i, 1, \frac{\lambda_i^2}{2} - 1) = \sum_i r_i$$

Analogously we decompose

$$s = (\sum \mu_i, N, \frac{(\sum \mu_i)^2}{2N} - 1) = \sum_i (\mu_i, 1, \frac{\mu_i^2}{2} - 1) = \sum s_i.$$

Now, $(r,s) = \sum_{i,j}(r_i, s_j)$. We will only describe the most important cases: a) the r_i, s_j form a diagram of type a_1^{2N}. Then all products r_i, s_j are zero and hence $(r,s) = 0$, leaving r and s unlinked, forming an $\mathbf{N}a_1^2$ (here the bold $\mathbf{N}$ is introduced to indicate long roots). b) the r_i, s_j form a diagram of the following type: for every i there is a unique j such that r_i and s_j are joined by a single bond. (The resulting diagram is of type a_2^N.) Adding up the inner products r_i, s_j shows that r and s will be joined by a single bond, thus forming $\mathbf{N}a_2$. It is now obvious how to identify the action of σ for some other constellations that may be needed. This completes the description of the induced action of σ on Dynkin diagrams.

The above results describe the relation of holes in the Leech lattice to those in $\mathcal{R}$. We will now proceed to develop some techniques which will help to identify the automorphism groups of generalized holes in $\mathcal{R}$. Throughout chapter 6, we will work with the normalizer of σ in Co_0 as the group of known automorphisms of the fixed point lattices. The normalizer (as in chapter 4.1, above) is the subgroup of Co_0 of automorphisms which normalize the cyclic group $\langle \sigma \rangle$. The normalizer of σ fixes Λ^σ. There may, or may not, exist further automorphisms of Λ^σ. The full automorphism group $\text{Aut}(\Lambda^\sigma)$ will be identified case by case in the course of the calculations of chapter 6.

THEOREM 5.9. *Suppose $\langle H \rangle$ is a generalized hole of $\mathcal{R}$ of type (I) or (II), as defined in theorem 5.8. Let $\langle H_\Lambda \rangle$ denote the unique hole of the Leech lattice associated with $\langle H \rangle$ in theorem 5.8. Suppose further that in the Leech lattice all holes $\langle K_\Lambda \rangle$ with $\Delta(K_\Lambda) = \Delta(H_\Lambda)$ are equivalent to $\langle H_\Lambda \rangle$ under the automorphism group Co_0 of Λ. Suppose that the order of $\text{Aut}(H_\Lambda)$ is divisible by N but not by N^2. Then all holes $\langle K \rangle$ in $\mathcal{R}$ with $\Delta(K) = \Delta(H)$ are equivalent to $\langle H \rangle$ under the automorphism group of Λ^σ.*

PROOF. Suppose $\langle K \rangle$ is another hole of $\mathcal{R}$ with $\Delta(K) = \Delta(H)$. Then there exists a unique corresponding hole $\langle K_\Lambda \rangle$ of the Leech lattice. $\Delta(K_\Lambda)$ must be equal to $\Delta(H_\Lambda)$ because, in cases (I) and (II), all vertices of the Leech lattice holes are preimages of those vertices in $\mathcal{R}$. (The considerations above showed how to fully reconstruct the Leech lattice hole.) Hence, there is a $\phi \in Co_0$ such that $K_\Lambda = \phi(H_\Lambda)$. σ acts on K_Λ. Consider $\phi^{-1}\sigma\phi$ and σ. Both are automorphisms of order N, acting on H_Λ. The cyclic groups generated by $\phi^{-1}\sigma\phi$ and σ must be conjugate in $\text{Aut}(H_\Lambda)$ because, by assumption, the order of this group is divisible by N, and not divisible by N^2 (see Sylow's theorem, e.g., [**Jac85**], p.80), say by some $\psi \in \text{Aut}(H_\Lambda)$. Then $\sigma^n = \psi^{-1}\phi^{-1}\sigma\phi\psi$ for some integer n, which implies that $\phi\psi$ is an element of the normalizer of σ. This shows that $\phi\psi$ is element of $\text{Aut}(\Lambda^\sigma)$. □

The above theorem will cover many of the occurring diagrams. However, the rest will have to be treated on a case by case basis, using the following technique:

THEOREM 5.10. *Suppose we consider two types of diagram, Δ_1, and Δ_2. They may be of any of the three types (I), (II), or (III) specified in theorem 5.8. Suppose there are n_1 generalized holes of type Δ_1, and n_2 generalized holes of type Δ_2, within a fundamental region of the fixed point lattice. Suppose it is known that all holes of type Δ_1 are equivalent under $Aut(\Lambda^\sigma)$. Suppose a hole $\langle H_1 \rangle$ such that $\Delta(H_1) = \Delta_1$ has f_1 faces F which border on holes $\langle H_2 \rangle$ of type $\Delta(H_2) = \Delta_2$ such that there exist diagram automorphisms in $Aut(H_1)$ acting transitively on the f_1 faces F. Suppose the neighbouring holes $\langle H_2 \rangle$ of type Δ_2 have f_2 faces of type F bordering on holes of type Δ_1. Suppose that $n_1 f_1 = n_2 f_2$. Then all diagrams of type Δ_2 are equivalent under $Aut(\Lambda^\sigma)$.*

PROOF. We use the automorphisms identifying the holes of type Δ_1, and those of $Aut(H_1)$ to identify all holes of type Δ_2 which border on holes of type Δ_1. The assumption $n_1 f_1 = n_2 f_2$ then implies that we have accounted for all diagrams of type Δ_2. $\square$

We conclude the chapter with a number of remarks. The check of the assumptions of theorem 5.10 will be simplified if $f_1 = 1$. We know from lemma 5.3 about the faces of holes. We understand the action of σ on Dynkin diagrams. Hence we can use theorem 5.9 to provide plenty of starting points so that some careful planning of the correct route through the remaining types of generalized holes will almost always succeed in selecting such neighbours. Finally, in chapter 6 we will first develop a strategy to count the numbers of holes of any type within a fundamental region of the fixed point lattice. The application of theorem 5.10 will provide a double check of the consistency of these numbers. This is all the more valuable as the complexity of the search means that it will only be possible to present the results, not however all individual calculations.

CHAPTER 6

Hyperbolic Lie Algebras

A Lie algebra is called hyperbolic if its Cartan matrix satisfies the condition that every proper principal minor is of finite or affine type. [**Wan91**] gave a complete list of all 238 hyperbolic Lie algebras with more than 2 simple roots. The significant difference to finite and affine Lie algebras is that hyperbolic Lie algebras possess roots of negative norm. Some of their multiplicities will be greater than 1. The purpose of this chapter is to identify the hyperbolic subalgebras of the GKMs $\mathcal{G}_N$ and thus to find upper bounds for the root multiplicities of these hyperbolic Lie algebras.

The project of finding hyperbolic subalgebras does not present any theoretical problems as we simply have to check all subsets (of correct size) of the set of those real simple roots in the GKM $\mathcal{G}_N$ such that the representing vectors in $\mathcal{R}$ have norm less or equal to 8 (as the root system $\mathcal{R}$ is translation invariant). However, in the case $N = 2$ this would require checking subsets of up to 10 roots of a set of several hundred thousand roots. No computer is fast enough to carry out such an undirected search. Hence we will devise a strategy for a more organized search.

As before, let σ be an automorphism of the Leech lattice of order N, and $N = 23, 11, 7, 5, 3, 2$. We consider the GKM $\mathcal{G}_N$ corresponding to σ as constructed in chapter 1. Section 6.1 recalls the classification of [**Wan91**] and identifies which hyperbolic Lie algebras are candidates to be subalgebras of one or any of the GKMs $\mathcal{G}_N$. Section 6.2 indicates how to carry out the explicit complete enumeration of all finite, affine, and hyperbolic, subalgebras of the GKMs $\mathcal{G}_N$ for all relevant N. The 2-dimensional case ($N = 23$, section 6.2.1) is almost trivial but very useful for visualizing the problem. The 4-dimensional case ($N = 11$, section 6.2.2) already shows all problems of the cases in higher dimensions. As the calculations become very tedious for the remaining four cases we will restrict ourselves in section 6.2.3.1 to an account of respective bases and symmetries used in the actual calculations. Section 6.2.3.2 provides a list of all those hyperbolic Lie algebras which can be identified as subalgebras of one of the GKMs $\mathcal{G}_N$ in such a way that the resulting upper bounds are sharp at least for some roots and such that they represent an improvement on existing bounds. Again, we can only give the results of the trivial but tedious calculations. For each such hyperbolic Lie algebra we give the Dynkin diagram and the root multiplicities for some imaginary roots of small height. I used Peterson's recursion formula ([**Kac90**], p.210) to carry out the numerical calculations of these root multiplicities. We contrast the multiplicities with the resulting upper bounds.

The complete classifications of finite and affine diagrams for all relevant N are provided in appendix A. Given that computer calculations were required to complete the proof, the function of sections 6.2.1 and 6.2.2 is not so much to carry out some trivial calculations but to show that the output of a computer program actually constitutes a mathematical proof. The partition of space as listed in appendix A is not merely auxiliary but a result in its own right as well and it extends the results obtained by Borcherds, Conway, and Queen for the Leech lattice (see [**CS88**], chapter 25).

The concluding section 6.3 tries to put into perspective the results obtained in the previous sections of chapter 6. In section 6.2 we observed that the root multiplicities of $\mathcal{G}_N$ provided good upper bounds for some hyperbolic Lie algebras while for others they did not bear any resemblance to the correct multiplicities. We use some well known results about the multiplicities of particularly the norm 0 vectors of hyperbolic Lie algebras to identify some conditions which are necessary to obtain sharp upper bounds. This is not simply a descriptive exercise but will provide us with a strategy which could help to determine sharp upper bounds for more hyperbolic Lie algebras. We will describe this strategy and relate it to some explicit numerical calculations of root multiplicities. We will further relate it to some results and conjectures of Borcherds. We will conclude with a number of observations which present open problems related to the results of this work.

6.1. Wan's classification

[**Wan91**] classified all hyperbolic Lie algebras. The notation introduced in chapter 5.2 was based on Wan's notation. We will continue using it here. To achieve the aims of this chapter we need to identify all hyperbolic Lie algebras which are subalgebras of the GKMs $\mathcal{G}_N$ constructed in chapter 1. Wan's classification provides the pool of our search. However, we cannot expect to find all hyperbolic Lie algebras as such subalgebras. For a start, all subalgebras of the GKMs $\mathcal{G}_N$ constructed in chapter 1 will be symmetrizable. Further, among the $\mathcal{G}_N$ we do not have any Lie algebra whose real simple roots have ratio of norms 4:1. Finally, from chapter 5.1 it follows that in any subalgebra of the GKMs $\mathcal{G}_N$ there are real simple roots of at most two different norms, as we consider prime N only. Hence we may restrict our attention to symmetrizable hyperbolic Lie algebras with simple roots whose norms have ratio 2:1, 3:1 only. This leaves us with a list of altogether 96 hyperbolic Lie algebras which are potential candidates.

6.2. Finite, Affine, and Hyperbolic Subalgebras

We now proceed to identify the finite, affine, and hyperbolic Lie subalgebras of the GKMs. We will present the cases $N = 23$ (section 6.2.1) and $N = 11$ (section 6.2.2) explicitly. The case $N = 23$ can be visualized easily as it is 2-dimensional. The case $N = 11$ already shows all essential features of the general case. Section 6.2.3 summarizes the results for the remaining N, that is $N = 7, 5, 3$, and 2.

6.2.1. N=23.

6.2.1.1. *Finite and Affine Subalgebras of N=23*. Let $\mathcal{G}_{23}$ denote the GKM constructed in chapter 1 from an automorphism σ of cycle shape $1^1 23^1$. Let Λ^σ be the 2-dimensional fixed point lattice and $L = \Lambda^\sigma \oplus II_{1,1}$ be the corresponding Lorentzian lattice. Its real simple roots r have been calculated in theorem 5.1 to be the norm 2 roots $\left(\lambda, 1, \frac{\lambda^2}{2} - 1\right)$ where $\lambda \in \Lambda^\sigma$ and the norm 46 roots $\left(\lambda, 23, \frac{\lambda^2}{46} - 1\right)$ where $\lambda \in 23(\Lambda^\sigma)^*$. A basis of Λ^σ is easily obtained as

$$\text{(6.1a)} \qquad \lambda_1 = \frac{1}{\sqrt{8}}(-3, \sqrt{23}), \qquad \lambda_2 = \frac{1}{\sqrt{8}}(5, \sqrt{23}).$$

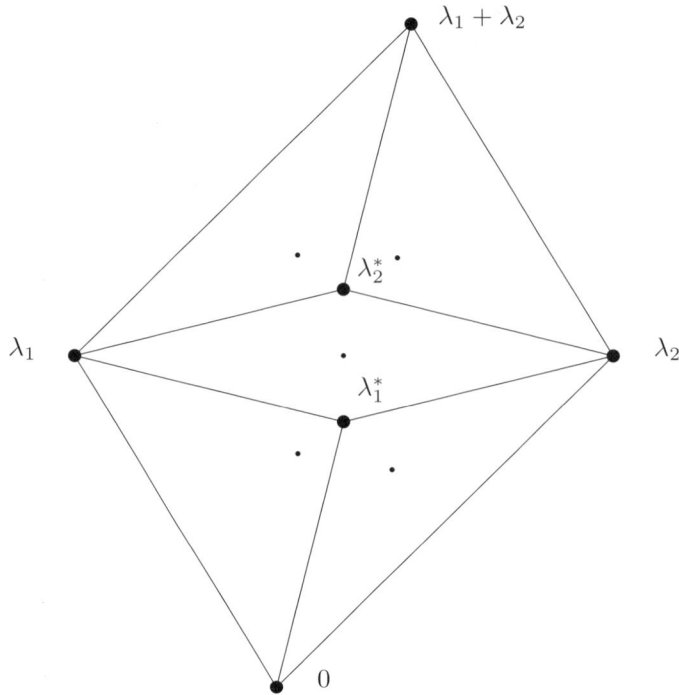

FIGURE 6.1. The root system of the GKM $\mathcal{G}_{23}$

We can therefore choose our fundamental region as the convex hull of the vectors 0, λ_1, λ_2, and $\lambda_1+\lambda_2$. We recall from chapter 4.1 (a remark following lemma 4.1) that the additive group $(\Lambda^\sigma)^*/\Lambda^\sigma$ has just one generator, which is of order 23. Hence, as a basis of the dual lattice we may choose

(6.1b) $$\frac{1}{23}(9\lambda_2 - 8\lambda_1) = \frac{1}{\sqrt{8}}(3, \frac{1}{23}\sqrt{23}), \qquad \lambda_1 - \lambda_2 = \frac{1}{\sqrt{8}}(8,0).$$

There are precisely 2 elements of $\mathcal{R}_{dual}$ within the fundamental region: $\lambda_1^* = \frac{1}{\sqrt{8}}(1, \frac{19}{23}\sqrt{23})$ (of norm $2+\frac{2}{23}$) and $\lambda_2^* = \frac{1}{\sqrt{8}}(1, \frac{27}{23}\sqrt{23})$ (of norm $4+\frac{2}{23}$). A diagram of the root system $\mathcal{R}$ for the fundamental region is shown in figure 6.1.

REMARK. The smaller dots indicate the positions of the generalized centres of the generalized holes in the partition. We further observe that the balls of radius $\sqrt{2}$ and centres at the fixed points $\mathcal{R}_{fix}$ do not cover the space. Correspondingly, the simple roots of norm 2 do not constitute the full set of real simple roots of the GKM $\mathcal{G}_{23}$.

The volume of the fundamental region is the determinant of the matrix formed by the vectors λ_1 and λ_2:

(6.2) $$V(23) = \det\begin{pmatrix} \frac{1}{\sqrt{8}}5 & \frac{1}{\sqrt{8}}\sqrt{23} \\ -\frac{1}{\sqrt{8}}3 & \frac{1}{\sqrt{8}}\sqrt{23} \end{pmatrix} = \sqrt{23}$$

Within the fundamental region we identify 5 generalized holes. There is one affine hole of type $A_1\ \mathbf{23}A_1$, two finite ones of type $a_1^2\ \mathbf{23}a_1$, and two finite ones of type $a_2\ \mathbf{23}a_1$. Using the automorphism $-\mathrm{id}$ of the fixed point lattice (and suitable translations by lattice elements) we observe that the finite holes of equal type are in fact equivalent. It is easily recognized that, in fact, the full automorphism group fixing 0 (see chapter 5.4) is of order 2. Formulae (5.14) to (5.17) of chapter 5.3 in conjunction with table 5.1 provide the volumes of these holes:

(6.3a) $$V(A_1\ \mathbf{23}A_1) = \frac{1}{2!}V(A_1)V(\mathbf{23}A_1) = \frac{1}{2!}(2\sqrt{2})(2\frac{\sqrt{2}}{\sqrt{23}}) = \frac{4}{\sqrt{23}}$$

(6.3b)
$$V(a_1^2\ \mathbf{23}a_1) = \frac{1}{2!}\sqrt{\left(\rho^2(a_1) + \rho^2(a_1) + \rho^2(\mathbf{23}a_1)\right)}$$
$$\times \sqrt{\left(\det(a_1)\det(a_1)\det(\mathbf{23}a_1)\right)}$$
$$= \frac{1}{2}\sqrt{\left(\frac{1}{2} + \frac{1}{2} + \frac{23}{2}\right)\left(2 \times 2 \times \frac{2}{23}\right)} = \frac{5}{\sqrt{23}}$$

(6.3c)
$$V(a_2\ \mathbf{23}a_1) = \frac{1}{2!}\sqrt{\left(\rho^2(a_2) + \rho^2(\mathbf{23}a_1)\right)\left(\det(a_2)\det(\mathbf{23}a_1)\right)}$$
$$= \frac{1}{2}\sqrt{\left(2 + \frac{23}{2}\right)\left(3 \times \frac{2}{23}\right)} = \frac{4.5}{\sqrt{23}}$$

The sum of volumes of the individual holes within any fundamental region hence works out to

(6.4) $$1 \times \frac{4}{\sqrt{23}} + 2 \times \frac{5}{\sqrt{23}} + 2 \times \frac{4.5}{\sqrt{23}} = \frac{23}{\sqrt{23}}$$

as expected from formula (6.2). This provides a check that the decomposition into generalized holes was both complete and error free.

6.2.1.2. *Hyperbolic Subalgebras for N=23.* We now search for hyperbolic Lie algebras contained in the root system of $\mathcal{G}_{23}$. As we do not have any affine, or finite, subalgebra containing more than 2 roots we can restrict ourselves to hyperbolic Lie algebras of rank 3. Furthermore, we will not obtain pointed arrows as the norm of the longer root is 46. We recall from chapter 5.2 that in the notation of [**Wan91**] a double arrow corresponds to a Cartan matrix $\begin{pmatrix} 2 & -2 \\ -2 & 2 \end{pmatrix}$, and that a single bond corresponds to a Cartan matrix $\begin{pmatrix} 2 & -1 \\ -1 & 2 \end{pmatrix}$. Checking [**Wan91**], we are left with the five potential candidates only which are shown in figure 6.2. All of these contain an affine subalgebra A_1. The decomposition of the volume has proved that up to automorphism there is a unique representative of A_1 within $\mathcal{R}$, which without loss of generality can be taken to be λ_1, λ_2. (Recall chapter 5.2, particularly formula (5.3), for details on how to translate Dynkin-diagrams into distances in $\mathcal{R}$.) We fix a double arrow within the first of the above diagrams. The rest of the diagram then translates into a pair of distances (4,6) of the remaining third root from λ_1 and λ_2. Clearly the 0 vector does satisfy these conditions. Hence we have found that the corresponding real simple roots of $\mathcal{G}_{23}$,

(6.5) $$\alpha_1 = (\lambda_1, 1, 1), \quad \alpha_2 = (\lambda_2, 1, 2)), \quad \alpha_3 = (0, 1, -1)$$

do represent the Dynkin diagram. This algebra is named $H_{96}^{(3)}$ in Wan's classification and AE_3 in [**Kac90**] (chapter 4). It is straightforward to check the remaining

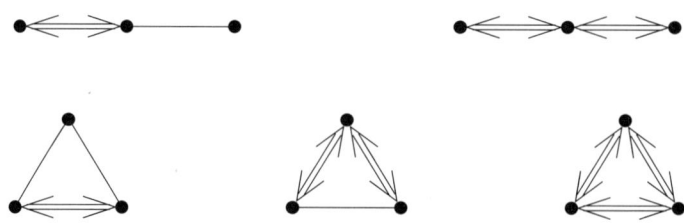

FIGURE 6.2. Candidates in the case $N = 23$

four candidates for hyperbolic subalgebras. They, however, cannot be realized within $\mathcal{R}$.

Let us return to AE_3. The corollary to theorem 1.7 now shows that

(6.6a) $$\mathrm{mult}(r) \leq p_\sigma(1 - \frac{r^2}{2}) \quad \text{if} \quad r \notin 23L^*$$

(6.6b) $$\mathrm{mult}(r) \leq p_\sigma(1 - \frac{r^2}{2}) + p_\sigma(1 - \frac{r^2}{46}) \quad \text{if} \quad r \in 23L^*.$$

Here $p_\sigma(1+n)$ is defined to be the coefficient of q^{1+n} in the q-expansion of $q\eta^{-1}(q)\eta^{-1}(q^{23})$ as in formula (1.24). We will want to compare these coefficients with some obtained in [**FF83**]. For convenience, we rewrite the definition of the p_σ as follows:

(6.7) $$\sum_n p_\sigma(n) q^n = \prod_n (1 - q^n)^{-1} (1 + q^{23} + 2q^{46} + \ldots).$$

The imaginary roots can be partially ordered by norm and height and are described as a linear combination of the simple roots. In the table 6.1 below, '2,2,1' represents the imaginary root $2\alpha_1 + 2\alpha_2 + 1\alpha_3$ with roots as labelled in the Dynkin diagram. We list a number of imaginary roots of small height and give their multiplicities. The multiplicities can be found in [**Kac90**], p.215. They are reproduced here for convenience. Obviously, it is sufficient to register the roots of a Weyl chamber. The column 'bound' shows the upper bounds as obtained in formula (6.6).

6.2. FINITE, AFFINE, AND HYPERBOLIC SUBALGEBRAS 77

$$H_{96}^{(3)} = AE_3 \subset \mathcal{G}_{23}$$

$\alpha_1 \quad \alpha_2 \quad \alpha_3$
●⇐⇒●———●

coefficients	norm	mult	bound	coefficients	norm	mult	bound
1, 1, 0	0	1	1	10, 10, 2	-32	297	297
2, 2, 1	-2	2	2	17, 17, 1	-32	297	297
3, 3, 1	-4	3	3	8, 9, 3	-34	385	385
3, 4, 2	-6	5	5	10, 11, 2	-34	385	385
4, 4, 1	-6	5	5	18, 18, 1	-34	385	385
4, 4, 2	-8	7	7	9, 9, 3	-36	490	490
5, 5, 1	-8	7	7	11, 11, 2	-36	490	490
4, 5, 2	-10	11	11	19, 19, 1	-36	490	490
6, 6, 1	-10	11	11	8, 9, 4	-38	627	627
5, 5, 2	-12	15	15	11, 12, 2	-38	626	627
7, 7, 1	-12	15	15	20, 20, 1	-38	627	627
5, 6, 2	-14	22	22	8, 10, 4	-40	792	792
8, 8, 1	-14	22	22	9, 9, 4	-40	792	792
5, 6, 3	-16	30	30	9, 10, 3	-40	792	792
6, 6, 2	-16	30	30	12, 12, 2	-40	791	792
9, 9, 1	-16	30	30	21, 21, 1	-40	792	792
6, 6, 3	-18	42	42	8, 10, 5	-42	1002	1002
6, 7, 2	-18	42	42	10, 10, 3	-42	1002	1002
10, 10, 1	-18	42	42	12, 13, 2	-42	1001	1002
7, 7, 2	-20	56	56	13, 13, 2	-44	1253	1256
11, 11, 1	-20	56	56	9, 10, 4	-46	1574	1576
6, 7, 3	-22	77	77	10, 11, 3	-46	1574	1576
7, 8, 2	-22	77	77	13, 14, 2	-46	1571	1576
12, 12, 1	-22	77	77	9, 10, 5	-48	1957	1960
6, 8, 4	-24	101	101	9, 11, 4	-48	1957	1960
7, 7, 3	-24	101	101	10, 10, 4	-48	1957	1960
8, 8, 2	-24	101	101	11, 11, 3	-48	1956	1960
13, 13, 1	-24	101	101	14, 14, 2	-48	1953	1960
8, 9, 2	-26	135	135	10, 10, 5	-50	2434	2439
14, 14, 1	-26	135	135	14, 15, 2	-50	2429	2439
7, 8, 3	-28	176	176	9, 11, 5	-52	3007	3015
9, 9, 2	-28	176	176	11, 12, 3	-52	3005	3015
15, 15, 1	-28	176	176	15, 15, 2	-52	3000	3015
7, 8, 4	-30	231	231	9, 12, 6	-54	3712	3725
8, 8, 3	-30	231	231	10, 11, 4	-54	3713	3725
9, 10, 2	-30	231	231	12, 12, 3	-54	3710	3725
16, 16, 1	-30	231	231	15, 16, 2	-54	3702	3725
7, 9, 4	-32	297	297	10, 12, 4	-56	4557	4576
8, 8, 4	-32	297	297	11, 11, 4	-56	4557	4576

TABLE 6.1. root multiplicities of AE_3

In the context of hyperbolic Lie algebras we require the definition of the level of a root of a hyperbolic Lie algebra. Suppose we are given a hyperbolic Lie algebra of, say, n simple roots. Suppose that its Dynkin diagram contains a unique affine subalgebra, which then necessarily has $n-1$ simple roots, determining a unique simple root α_0 not in the affine diagram. If we express any root r as a positive linear combination of the simple roots $r = \sum n_i \alpha_i$ the level of r is defined to be the coefficient n_0 of α_0. Hence in our labelling of the simple roots the level of a root of AE_3 is simply the coefficient of α_3. It follows from proposition 10.10 and exercise 11.7 of [**Kac90**] that for simply laced hyperbolic Lie algebras all roots of level 1 and rank n satisfy

$$(6.8) \qquad \mathrm{mult}(r) = p_{n-2}(1 - \frac{r^2}{2})$$

Here, $p_n(x)$ is the number of partitions of x into parts of n colours.

Let us return to AE_3. This algebra has been studied extensively in [**FF83**]. [**FF83**] define the following series $p(n)$ and $p'(n)$:

$$(6.9\mathrm{a}) \qquad \sum_n p(n) q^n = \prod_n (1-q^n)^{-1}$$

Thus $p(n)$ is the number of partitions of n.

$$(6.9\mathrm{b}) \qquad \sum_n p'(n) q^n = \left(\prod_n (1-q^n)^{-1}\right)\left(1 - q^{20} + q^{22} \pm \dots\right)$$

[**FF83**] then restate the root multiplicities for level 1 explicitly as follows:

$$(6.10\mathrm{a}) \qquad \mathrm{mult}(r) = p(1 - \frac{r^2}{2})$$

[**FF83**] is mainly concerned with level 2. For roots of level 2 they obtain:

$$(6.10\mathrm{b}) \qquad \mathrm{mult}(r) = p'(1 - \frac{r^2}{2})$$

Let us consider a root of level r_3, say

$$r = r_1 \alpha_1 + r_2 \alpha_2 + r_3 \alpha_3 = (r_1 \lambda_1 + r_2 \lambda_2, r_1 + r_2 + r_3, r_1 + 2r_2 - r_3).$$

This will be an element of $23L^*$ if and only if

(6.11a) $\qquad\qquad r_1 \lambda_1 + r_2 \lambda_2 \in 23(\Lambda^\sigma)^*$

(6.11b) $\qquad\qquad r_1 + r_2 + r_3 \equiv 0(23)$

(6.11c) $\qquad\qquad r_1 + 2r_2 - r_3 \equiv 0(23)$

Without loss of generality we can consider coefficients r_1, r_2, r_3 modulo 23. Comparing with the basis (formula (6.1b)) of $(\Lambda^\sigma)^*$, we conclude that (6.11a) will only be satisfied if $r_1 \equiv -8n$, $r_2 \equiv 9n$, for some integer n. We feed this into (6.11b) and obtain $r_3 \equiv -n$. Now, (6.11c) reads $11n \equiv 0$. Thus, in summary, r will be an element of $23L^*$ if and only if all three coefficients r_1, r_2, r_3 are divisible by 23.

In particular, we observe that, for r of level 1 or 2, $r \notin 23L^*$. Hence, for these r the upper bounds of (6.6a) apply. We can now compare our upper bound (6.7) to the exact values (6.9a) and (6.9b). We find that our results are very close to the exact multiplicities where these are known. At the same time, our results apply to all roots of AE_3.

Furthermore, we observe that the five real simple roots $0, \lambda_1, \lambda_2, \lambda_1^*, \lambda_2^*$ form a symmetrized Cartan matrix as follows:

$$\tilde{C} = \begin{pmatrix} 2 & -2 & 0 & 0 & 0 \\ -2 & 2 & -1 & 0 & 0 \\ 0 & -1 & 2 & -23 & 0 \\ 0 & 0 & -23 & 46 & -46 \\ 0 & 0 & 0 & -46 & 46 \end{pmatrix}$$

In particular, the equation $23(\lambda_1 + \lambda_2) = \lambda_1^* + \lambda_2^*$ shows that for every root in $23L^*$ there are additional roots in $\mathcal{G}_{23}$ which are not in the hyperbolic Lie algebra AE_3. This provides some idea why the upper bounds are not exact for all roots.

6.2.2. N=11.

6.2.2.1. *Finite and Affine Subalgebras.* Let $\mathcal{G}_{11}$ denote the GKM constructed in chapter 1 from an automorphism σ of cycle shape $1^2 11^2$. Let Λ^σ be the 4-dimensional fixed point lattice and $L = \Lambda^\sigma \oplus II_{1,1}$ be the corresponding Lorentzian lattice. The set of its real simple roots $\mathcal{R}$ has been calculated in theorem 5.1. A basis of Λ^σ is easily obtained as

$$(-3, 1, \sqrt{11}, \sqrt{11}), \quad (2, 0, 2\sqrt{11}, 0), \quad (4, 4, 0, 0), \quad (8, 0, 0, 0).$$

Here, as in the rest of this section, we operate in basic units of $\frac{1}{\sqrt{8}}$, that is, we suppress this factor, which is common to all vectors considered. The basis vectors form a diagonal matrix and we can read off the fundamental volume to be $\sqrt{11}^2$.

The fixed point lattice is a sublattice of the Leech lattice. Hence the shortest vectors will have norm 4. From the considerations of chapter 5.2, formula (5.3) it follows that any finite or affine diagram has bonds which correspond to distances between the roots either 4, or 6, or 8. Without loss of generality we can choose the 0 vector as one of the roots. Then we can restrict our attention to vectors of norm 4, 6, and 8. A complete list of these is easily obtained from the explicit basis and given in table 6.2:

norm 4 vectors	norm 6 vectors	norm 8 vectors
$(4, 4, 0, 0)$	$(2, 0, 2\sqrt{11}, 0)$	$(8, 0, 0, 0)$
$(4, -4, 0, 0)$	$(0, 2, 0, 2\sqrt{11})$	$(0, 8, 0, 0)$
$(3, -1, -\sqrt{11}, -\sqrt{11})$	$(5, 1, \sqrt{11}, \sqrt{11})$	$(4, -2, 0, 2\sqrt{11})$
$(-1, 3, -\sqrt{11}, -\sqrt{11})$	$(1, 5, \sqrt{11}, \sqrt{11})$	$(-4, -2, 0, 2\sqrt{11})$
$(3, 1, -\sqrt{11}, \sqrt{11})$	$(5, -1, \sqrt{11}, -\sqrt{11})$	$(-2, 4, 2\sqrt{11}, 0)$
$(1, 3, -\sqrt{11}, \sqrt{11})$	$(-1, 5, -\sqrt{11}, \sqrt{11})$	$(-2, -4, 2\sqrt{11}, 0)$

TABLE 6.2. Short vectors of Λ^σ

We now turn to the elements of $\mathcal{R}_{dual}$. As established in chapter 5.2 (remark following theorem 5.4), there cannot be any bonds between elements of $\mathcal{R}_{fix}$ and $\mathcal{R}_{dual}$ in a diagram of finite or affine type. If we require 0 to be one of the roots we can restrict our attention to those elements of the dual which have norm exactly $2 + \frac{2}{11}$. To identify those, we begin by recalling the remark following lemma 4.1

that the additive group $(\Lambda^\sigma)^*/\Lambda^\sigma$ has two generators, each of order 11. We can choose them as

$$[\lambda_1^*] = (3, -1, \frac{\sqrt{11}}{11}, \frac{\sqrt{11}}{11}), \qquad [\lambda_2^*] = (3, 1, \frac{\sqrt{11}}{11}, \frac{-\sqrt{11}}{11}).$$

To find elements of the dual lattice of norm $2 + \frac{2}{11}$ we require $(a[\lambda_1^*] + b[\lambda_2^*])^2 \equiv \frac{2}{11}$ mod $2\mathbb{Z}$, where a, b are integers modulo 11. This is equivalent to the condition

(6.12) $$a^2 + b^2 \equiv 8 \bmod 11.$$

As established in lemma 4.2, this identity has 12 solutions: They are $\{(\pm 2, \pm 2), (\pm 4, \pm 5), (\pm 5, \pm 4)\}$. We consider the solution $(2, 2)$. The corresponding equivalence class is represented by the vector $(4, 0, \frac{4}{11}\sqrt{11}, 0)$. A check of all short vectors establishes that there are precisely 6 representatives of norm $2 + \frac{2}{11}$, namely $(\pm 4, 0, \frac{4}{11}\sqrt{11}, 0)$, $(0, \pm 4, \frac{4}{11}\sqrt{11}, 0)$, $(-1, \pm 1, \frac{-7}{11}\sqrt{11}, \frac{\pm 11}{11}\sqrt{11})$.

We recall the claim, (sketch) proved and used in the deduction of theorem 4.2, that $GO_2(11)$ acts transitively on the 12 solutions of (6.12). It furthermore lifts to automorphisms of the fixed point lattice. Hence we have found that there are $12 \times 6 = 72$ elements of norm $2 + \frac{2}{11}$ within $\mathcal{R}_{dual}$. From here, it is straightforward to calculate their co-ordinates. We will, however, omit a complete list.

The search for generalized holes in $\mathcal{R}$ will be simplified if we can make use of the automorphisms of the lattice. There are a number of obvious symmetries of order 2:

(6.13) $$\phi_1 = -\mathrm{id} \qquad \phi_2 = \begin{pmatrix} 1 & 0 & 0 & 0 \\ 0 & -1 & 0 & 0 \\ 0 & 0 & 1 & 0 \\ 0 & 0 & 0 & -1 \end{pmatrix} \qquad \phi_3 = \begin{pmatrix} 0 & 1 & 0 & 0 \\ 1 & 0 & 0 & 0 \\ 0 & 0 & 0 & 1 \\ 0 & 0 & 1 & 0 \end{pmatrix}$$

Referring to the complete list of short vectors we observe that there are just two types of norm 8 vectors which cannot be identified by the above automorphisms. As the centralizer of σ in Co_0 has order 66 we may now conjecture that there exists an automorphism of order 3 identifying the two types. (It must be stressed that it can only be a pious hope to expect that all norm 8 vectors will be equivalent, even though it is true for Leech lattice vectors. It is, for example, false in the case $N = 2$, where there exist two types of diagrams, A_1^{16} and $A_1^8\, \mathbf{2}A_1^8$.) Let us try to construct an automorphism ϕ mapping $(8, 0, 0, 0)^T$ to $(4, -2, 0, 2\sqrt{11})^T$. Thus the first column of ϕ will be $(\frac{1}{2}, -\frac{1}{4}, 0, \frac{1}{4}\sqrt{11})^T$. $(0, 8, 0, 0)^T$ must now be mapped to some other norm 8 vector which still will be orthogonal to $\phi(8, 0, 0, 0)^T$. Checking the list of norm 8 vectors there are precisely 2 possible choices: $\pm(2, 4, -2\sqrt{11}, 0)^T$. (These two choices are equivalent because of the automorphism ϕ_2, formula (6.13)). Continuing in this way we obtain

(6.14) $$\phi = \begin{pmatrix} \frac{1}{2} & \frac{1}{4} & 0 & -\frac{1}{4}\sqrt{11} \\ -\frac{1}{4} & \frac{1}{2} & \frac{1}{4}\sqrt{11} & 0 \\ 0 & -\frac{1}{4}\sqrt{11} & \frac{1}{2} & -\frac{1}{4} \\ \frac{1}{4}\sqrt{11} & 0 & \frac{1}{4} & \frac{1}{2} \end{pmatrix}$$

We observe that ϕ does not only identify the two types of norm 8 vectors which were not equivalent under the automorphism group generated by ϕ_1, ϕ_2, ϕ_3 but it does so for the norm 4 and norm 6 vectors as well.

6.2. FINITE, AFFINE, AND HYPERBOLIC SUBALGEBRAS

We begin the decomposition of space into generalized holes by searching for A_1 diagrams, the only type requiring norm 8 vectors. As all norm 8 vectors are equivalent under the automorphism group we choose the vectors 0 and $(8,0,0,0)$ without loss of generality. This is a complete affine component, hence all remaining elements of this hole must have minimal distance from both. This obviously leaves the vectors $(4, \pm 4, 0, 0)$, and the dual vectors $(4, 0, \pm \frac{4}{11}\sqrt{11}, 0)$, and $(4, 0, \pm \frac{4}{11}\sqrt{11}, 0)$. Thus we have identified an $A_1^2\, \mathbf{11}A_1^2$. Now there are 12 norm 8 vectors which each form one such diagram with 0. On the other hand, within each diagram there are 4 ways of placing the 0 and norm 8 vectors. Hence, within any fundamental volume, there are $\frac{12}{4} = 3$ such diagrams.

Next, let us search for diagrams of the type $a_1^n\, \mathbf{11}\Delta$ where Δ is any (not necessarily undecomposable) simply laced diagram of finite type. Theorem 5.8 asserts that among the vertices of any generalized hole there are at least $M+1$ elements of $\mathcal{R}_{fix}$. Hence n will be greater or equal 3. Without loss of generality, choose 0 and any of the norm 4 vectors, say $(4,4,0,0)$. There are only 2 vectors left which have distance 4 to both the above: $(3, 1, -\sqrt{11}, \sqrt{11})$ and $(1, 3, \sqrt{11}, -\sqrt{11})$. Automorphism ϕ_3 shows that the 2 choices for a third vector are equivalent. Furthermore, we observe that there cannot be an a_1^4 diagram. We choose, say, $(3, 1, -\sqrt{11}, \sqrt{11})$. Of the 72 dual vectors there are precisely 6 left which have minimal distance from the 3 chosen vectors. We refer to them as vectors a to f, respectively. We calculate their distance table, where a '*' stands in the place of any distance greater $\frac{6}{11}$ as such a distance cannot occur within a finite diagram. The distances are given in units of $\frac{1}{11}$:

	a	b	c	d	e	f
a		4	*	*	*	*
b			*	*	*	*
c				*	*	4
d					4	*
e						*

Hence, we identify three diagrams of type $a_1^3\, \mathbf{11}a_1^2$. We have had 12 choices of the first norm 4 vector, another 2 choices for the second. On the other hand, there are 3! ways of choosing which of the vectors corresponds to which of the a_1^3. Hence, within any fundamental volume, there are $\frac{12 \times 2 \times 3}{3!} = 12$ holes of type $a_1^3\, \mathbf{11}a_1^2$.

Any generalized hole which is not of either of the above two types will contain a single bond, that is a pair of vectors of $\mathcal{R}_{fix}$ at distance 6. We choose without loss of generality 0 and $(5, 1, \sqrt{11}, \sqrt{11})$. We first complete the search for affine diagrams. There are only 2 affine holes in the Leech lattice which admit an automorphism of order 11. They are A_1^{24} and A_2^{12}. Theorem 5.8 (I) asserts that all generalized affine holes of $\mathcal{R}$ lift to affine holes of the Leech lattice whose automorphism group has order divisible by N. Hence we can restrict the affine search to A_1 and A_2. There are exactly 2 vectors of norm 6 which have distance 6 from $(5, 1, \sqrt{11}, \sqrt{11})$. They are $(5, -1, \sqrt{11}, -\sqrt{11})$ and $(2, 0, 2\sqrt{11}, 0)$. Obviously these two choices to form an A_2 are equivalent as the respective triplets of vectors are translations of one another. Choosing either, and collecting all dual vectors that have minimal distance to the three vectors we obtain a unique diagram $A_2\, \mathbf{11}A_2$. Taking into consideration the choices made we found $\frac{12 \times 2}{3 \times 2} = 4$ such diagrams within the fundamental volume.

We now turn to finite diagrams. Any such contains roots which are joined only to one other root. Without loss of generality we can assume that the 0 root is chosen to be such. Then, apart from the fixed choice of one norm 6 vector, we will only have to consider norm 4 vectors which have distance less or equal to 6 from the chosen vector of norm 6. (This may not seem important, however, it does make a difference if there are 4320 vectors of norm 4 and 61440 of norm 6, as in the case $N = 2$.) In the case at hand, there are 10 elements of $\mathcal{R}_{dual}$ of minimal distance to both 0 and $(5, 1, \sqrt{11}, \sqrt{11})$. We will refer to them as a to j, respectively. The distance table (in units of $\frac{1}{11}$, as above) is as follows:

	a	b	c	d	e	f	g	h	i	j
a		4	*	*	6	6	*	*	*	*
b			*	*	*	*	*	*	6	*
c				4	*	*	6	*	*	6
d					*	*	*	6	*	*
e						6	*	*	*	*
f							*	4	*	*
g								*	*	6
h									*	*
i										4

We list the norm 4 vectors which satisfy all conditions imposed so far. Let 'dist' denote the distance to $(5, 1, \sqrt{11}, \sqrt{11})$. 'duals' lists those elements of $\mathcal{R}_{dual}$ which have minimal distance from the respective norm 4 vector:

$$A = (4, 4, 0, 0) \quad \text{dist} = 4 \quad \text{duals } a, b, f, g, h$$
$$B = (4, -4, 0, 0) \quad \text{dist} = 6 \quad \text{duals } a, b, e, i$$
$$C = (1, -3, \sqrt{11}, \sqrt{11}) \quad \text{dist} = 4 \quad \text{duals } c, d, e, i, j$$
$$D = (3, 1, -\sqrt{11}, \sqrt{11}) \quad \text{dist} = 6 \quad \text{duals } b, g, i, j$$

Every generalized hole of finite type in the case $N = 11$ has 5 vertices. We begin by searching for diagrams that contain the minimal number of elements of $\mathcal{R}_{fix}$, which is 3 (theorem 5.8). We then need 2 elements of $\mathcal{R}_{dual}$ to complete the diagram. Choose A. ab, and fh, yield $a_2a_1\,\mathbf{11}a_1^2$. af yields $a_2a_1\,\mathbf{11}a_2$. Choose C. cd, and ij, yield $a_2a_1\,\mathbf{11}a_1^2$. cj yields $a_2a_1\,\mathbf{11}a_2$. Choose B. ab yields $a_3\,\mathbf{11}a_1^2$. ae, and bi, yield $a_3\,\mathbf{11}a_2$. Choose D. ij yields $a_3\,\mathbf{11}a_1^2$. bi, and gj, yield $a_3\,\mathbf{11}a_2$. Next, let us search for diagrams containing 4 elements of $\mathcal{R}_{fix}$ and 1 element of $\mathcal{R}_{dual}$. Choose AD. b, and g, yield $a_3a_1\,\mathbf{11}a_1$. Choose BC. e, and i, yield $a_3a_1\,\mathbf{11}a_1$. There are no further diagrams containing 4 elements of $\mathcal{R}_{fix}$ because the distances AB, AC, BD are greater than 6. For the same reason, there can be no diagrams containing 5 elements of $\mathcal{R}_{fix}$. In conclusion, we have obtained:

$$\text{type } a_3\,\mathbf{11}a_1^2: \quad \frac{12 \times (1+1)}{2} = 12 \text{ copies.}$$
$$\text{type } a_3\,\mathbf{11}a_2: \quad \frac{12 \times (2+2)}{2} = 24 \text{ copies.}$$
$$\text{type } a_2a_1\,\mathbf{11}a_1^2: \quad \frac{12 \times (2+2)}{2} = 24 \text{ copies.}$$
$$\text{type } a_2a_1\,\mathbf{11}a_2: \quad \frac{12 \times (1+1)}{2} = 12 \text{ copies.}$$
$$\text{type } a_3a_1\,\mathbf{11}a_1: \quad \frac{12 \times (2+2)}{2} = 24 \text{ copies.}$$

We can now proceed to calculate the volumes of the individual holes and carry out the volume check, just as in section 6.2.1 for the case $N = 23$. The details for the case $N = 11$ can be found in appendix A.

So far we have established the numbers of holes of the various types. We now turn to the question whether all holes of same type are equivalent under the automorphism group of the fixed point lattice. We know that 11^2 does not divide the order of Co_0, thus certainly not the order of any automorphism group of a hole

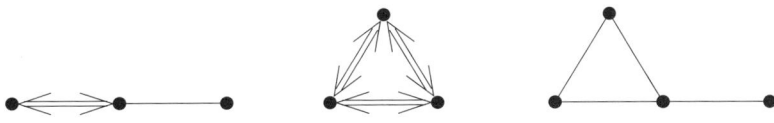

FIGURE 6.3. Hyperbolic Subalgebras of $\mathcal{G}_{11}$

in the Leech lattice. Hence, theorem 5.9 covers all affine diagrams and all finite diagrams which contain precisely 3 elements of $\mathcal{R}_{fix}$. This only leaves the diagram $a_3 a_1 \, \mathbf{11} a_1$ to be investigated. We consider a diagram of type $a_3 \, \mathbf{11} a_2$. This has two faces of type $a_3 \, \mathbf{11} a_1$. If we consider a representative of this type we will discover that the adjacent holes are $a_3 \, \mathbf{11} a_1^2$ and $a_3 a_1 \, \mathbf{11} a_1$, respectively. Hence, the conditions of theorem 5.10 are satisfied and we conclude that there are 24 holes of type $a_3 a_1 \, \mathbf{11} a_1$ equivalent under the automorphism group. This concludes the argument as there are precisely 24 such copies within any fundamental region.

We now turn to the size of the automorphism group. Again, we consider the diagram $a_3 \, \mathbf{11} a_2$. We established already that its two faces of type $a_3 \, \mathbf{11} a_1$ are not symmetric. But so are the two faces of type $a_2 \, \mathbf{11} a_2$, the adjacent diagrams being $A_2 \, \mathbf{11} A_2$ and $a_2 a_1 \, \mathbf{11} a_2$. Hence, the hole $a_3 \, \mathbf{11} a_2$ will only be preserved by the identity automorphism. As there exist 24 copies of it within a fundamental volume, the total automorphism group must have order 24. Thus we have verified the complete decomposition of the fundamental region as given in appendix A.

6.2.2.2. *Hyperbolic Subalgebras.* We can now carry out the search for hyperbolic subalgebras. We observe that there are no finite, or affine, subalgebras of more than 3 roots. Hence, we can restrict ourselves to hyperbolic Lie algebras of rank 3 and 4, whose roots, as in the case $N = 23$, must all be of equal length. That leaves us with 5 hyperbolic Lie algebras of rank 3 and 3 of rank 4 from the list of [**Wan91**]. They contain an A_1 or A_2 subdiagram, respectively. From the decomposition of space in section 6.2.2.1 we know that up to isomorphism there are two unique representatives of these in $\mathcal{R}$, one consisting of roots of norm 2, the other consisting of roots of norm $2N$. All that remains now is to check the elements of $\mathcal{R}$ close to the representing diagrams A_1, A_2, $\mathbf{11}A_1$, and $\mathbf{11}A_2$. We find that there exist exactly 3 hyperbolic Lie subalgebras of $\mathcal{G}_{11}$, shown in figure 6.3.

The first is the algebra AE_3 which we identified as a subalgebra of $\mathcal{G}_{23}$. As the upper bounds arising from $N = 11$ are worse than those arising from $N = 23$ we can discard this case. The second can be realized by the vectors 0, $(8, 0, 0, 0)$, and $(4, 2, 0, -2\sqrt{11})$. The third is represented through the vectors 0, $(5, 1, \sqrt{11}, \sqrt{11})$, $(5, -1, \sqrt{11}, -\sqrt{11})$, and $(4, 4, 0, 0)$. We do not find any additional hyperbolic subalgebras consisting of long roots.

As in the case $N = 23$, we obtain upper bounds for the root multiplicities of the above hyperbolic Lie algebras. We proceed to compare these bounds with the exact values for some imaginary roots, ordered by norm and height. The notation is as before (see section 6.2.1.2), and I calculated the exact root multiplicities by a program based on the Peterson recursion formula for root multiplicities (see [**Kac90**], p.210).

6. HYPERBOLIC LIE ALGEBRAS

$H_{71}^{(3)} \subset \mathcal{G}_{11}$

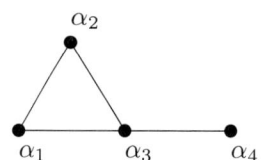

N=11

root-coefficients	norm	mult	bound
1, 1, 0	0	1	2
1, 0, 1	0	1	2
0, 1, 1	0	1	2
1, 1, 1	-6	2	20
2, 1, 1	-8	3	36
1, 2, 1	-8	3	36
1, 1, 2	-8	3	36
2, 2, 1	-14	6	185

$H_3^{(4)} = AE_4 \subset \mathcal{G}_{11}$

N=11

root-coefficients	norm	mult	bound
1, 1, 1, 0	0	2	2
2, 2, 2, 1	-2	5	5
3, 3, 3, 1	-4	10	10
3, 3, 4, 2	-6	20	20
4, 4, 4, 1	-6	20	20
4, 4, 4, 2	-8	36	36
5, 5, 5, 1	-8	36	36
4, 4, 5, 2	-10	65	65
6, 6, 6, 1	-10	65	65
5, 5, 5, 2	-12	110	110
5, 4, 6, 3	-12	110	110
4, 5, 6, 3	-12	110	110
7, 7, 7, 1	-12	110	110
5, 5, 6, 2	-14	185	185
8, 8, 8, 1	-14	185	185
5, 5, 6, 3	-16	300	300
6, 6, 6, 2	-16	300	300
6, 6, 7, 2	-18	481	481
6, 6, 6, 3	-18	481	481
6, 5, 7, 3	-18	481	481
5, 6, 7, 3	-18	481	481
7, 7, 7, 2	-20	752	754
6, 6, 7, 3	-22	1165	1169
7, 7, 8, 2	-22	1164	1169
7, 7, 7, 3	-24	1770	1780
7, 6, 8, 3	-24	1769	1780
6, 7, 8, 3	-24	1769	1780
6, 6, 8, 4	-24	1769	1780
8, 8, 8, 2	-24	1767	1780
7, 6, 8, 4	-26	2663	2685
6, 7, 8, 4	-26	2663	2685
7, 7, 8, 3	-28	3950	3996
7, 7, 8, 4	-30	5812	5894

TABLE 6.3. Root Multiplicities of Subalgebras of $\mathcal{G}_{11}$

We observe that the upper bounds for AE_4 are again very useful, though for roots of larger height there do occur discrepancies between the correct values and the upper bounds, similarly to the results for AE_3. However, the results for $H_{71}^{(3)}$ are so poor that they must be regarded as useless. In fact, the general bounds provided by theorem 1.4b are actually far closer than our result. We will return to this in section 6.3.

6.2.3. The remaining N.

6.2.3.1. *Finite and Hyperbolic Subalgebras.* The previous section showed in detail how to analyse the root system $\mathcal{R}$ of $\mathcal{G}_{11}$. I carried out the analysis of $\mathcal{G}_N$ for $N = 7, 5, 3, 2$ using the same principal line of the argument in all these cases. The numbers of vectors of norms 4 and 6 do, however, increase to levels which are better treated by computers. Again, it was necessary to make extensive use of the known symmetries of the fixed point lattice in order to reduce the number of individual representatives of each type that had to be identified and counted. I determined the symmetries by explicit inspection. Given an understanding of the symmetries, I was then in the position to carry out a computerized search for generalized holes, using preselections similar to those of A_1, a_1^n, a_2 in the case $N = 11$ demonstrated in section 6.2.2.1, above. Later, I transformed the numbers of representatives found into numbers per fundamental region and carried out the volume check manually. I subsequently analysed the automorphism groups of the individual holes by inspection (with computerized search of representing holes). This provided an important double check for the results. The checks of the automorphism groups yielded the following result: In the cases $N = 2, 5, 7, 11, 23$ it is true that, whenever two generalized holes have the same Dynkin diagram, they are equivalent under the automorphism group of Λ^σ. However, in the case $N = 3$, there do exist two pairs of generalized holes which represent the same Dynkin diagram but are inequivalent under the automorphism group of Λ^σ. One of these pairs concerns the Dynkin diagram a_8 $\mathbf{3}a_5$ (of type III of theorem 5.8), the other the diagram $a_5 a_2$ $\mathbf{3}a_2^3$ (of type II of theorem 5.8). This result implies that the theoretical analysis of the holes as carried out in section 5.4 cannot be carried much further. In particular, one cannot infer the equivalence of two holes in Λ^σ from the equivalence of their lifts in Λ, unless the automorphism group of the hole in Λ satisfies the assumptions of theorem 5.9.

In this section we restrict ourselves to listing the bases of the fixed point lattices from which we can read off the volume of a fundamental region. The complete decompositions of the fundamental regions are given in appendix A.

For $N = 7$ we choose as basis of the fixed point lattice (in units of $\frac{1}{\sqrt{8}}$):

$$\begin{pmatrix} -3 \\ 1 \\ 1 \\ \sqrt{7} \\ \sqrt{7} \\ \sqrt{7} \end{pmatrix}, \begin{pmatrix} 0 \\ 2 \\ 0 \\ 0 \\ 2\sqrt{7} \\ 0 \end{pmatrix}, \begin{pmatrix} 2 \\ 0 \\ 0 \\ 2\sqrt{7} \\ 0 \\ 0 \end{pmatrix}, \begin{pmatrix} 4 \\ 0 \\ 4 \\ 0 \\ 0 \\ 0 \end{pmatrix}, \begin{pmatrix} 4 \\ 4 \\ 0 \\ 0 \\ 0 \\ 0 \end{pmatrix}, \begin{pmatrix} 8 \\ 0 \\ 0 \\ 0 \\ 0 \\ 0 \end{pmatrix}$$

The fundamental volume thus is $\sqrt{7}^3$. If we check the table of results we can confirm the size of the total automorphism group easily: The automorphism group of A_6 has size at most 14, as this already accounts for all possible moves of the (isolated) diagram. Hence the total group cannot be bigger than 1176. On the other hand, there are holes of automorphism group 1.

For $N = 5$ we choose as basis of the fixed point lattice (in units of $\frac{1}{\sqrt{8}}$):

$$\begin{pmatrix} -3 \\ 1 \\ 1 \\ 1 \\ \sqrt{5} \\ \sqrt{5} \\ \sqrt{5} \\ \sqrt{5} \end{pmatrix}, \begin{pmatrix} 2 \\ 2 \\ 0 \\ 2 \\ 0 \\ 0 \\ 2\sqrt{5} \\ 0 \end{pmatrix}, \begin{pmatrix} 2 \\ 0 \\ 2 \\ 2 \\ 0 \\ 2\sqrt{5} \\ 0 \\ 0 \end{pmatrix}, \begin{pmatrix} 0 \\ 2 \\ 2 \\ 2 \\ 2\sqrt{5} \\ 0 \\ 0 \\ 0 \end{pmatrix}, \begin{pmatrix} 4 \\ 0 \\ 0 \\ 4 \\ 0 \\ 0 \\ 0 \\ 0 \end{pmatrix}, \begin{pmatrix} 4 \\ 0 \\ 4 \\ 0 \\ 0 \\ 0 \\ 0 \\ 0 \end{pmatrix}, \begin{pmatrix} 4 \\ 4 \\ 0 \\ 0 \\ 0 \\ 0 \\ 0 \\ 0 \end{pmatrix}, \begin{pmatrix} 8 \\ 0 \\ 0 \\ 0 \\ 0 \\ 0 \\ 0 \\ 0 \end{pmatrix}$$

The fundamental volume thus is $\sqrt{5}^4$. If we check the table of results we can confirm the size of the total automorphism group easily: The automorphism group of A_4^2 has size at most 200, as this already accounts for all possible moves of the (isolated) diagram. Hence the total group cannot be bigger than 14400. On the other hand, there are holes of automorphism group 1.

The fixed point lattices in the cases $N = 3$ and $N = 2$ are well known to be K_{12} and Λ_{16}, respectively. Hence there is no need to give an explicit basis here. The fundamental volumes and total automorphism groups can be quoted from [**CS88**], chapter 4. The volumes are $\sqrt{3}^6$, and $\sqrt{2}^8$ respectively. (Observe that the determinant referred to in [**CS88**] is the square of the volume of interest here.)

6.2.3.2. *Hyperbolic Subalgebras.* Before we continue to list examples of hyperbolic Lie algebras we need to reflect on the results obtained so far. We observed that the bounds are sharp only for some cases. In the remainder of this section we will restrict ourselves to those hyperbolic Lie algebras for which the upper bounds bear some resemblance to the true multiplicities and at the same time represent an improvement on known upper bounds. In section 6.3 we will return in more detail to the question which cases are successful.

To completely analyse all the GKMs $\mathcal{G}_N$ and to determine all hyperbolic subalgebras again requires tedious calculations such that we can only report the results. However, we do not print a classification of all such hyperbolic subalgebras as, on the one hand, there is no obvious use for such a classification, on the other hand, given the classification of all finite and affine subalgebras as in appendix A, the classification of all hyperbolic subalgebras is a straightforward exercise. We restrict ourselves to recording the following experimental fact (though I cannot offer any explanation): in no $\mathcal{G}_N$ there exist hyperbolic subalgebras whose rank oversteps the value given by the following dimension formula,

$$\text{rank} = \frac{\dim \Lambda^\sigma}{2} + 2$$

even though there do exist affine subalgebras of higher rank (such as A_8 in $N = 3$, A_6 in $N = 7$, and D_6 in $N = 5$).

This means that hyperbolic Lie algebras of rank 7 or 8 have only been identified, if at all, as subalgebras of $\mathcal{G}_3$ and $\mathcal{G}_2$, hyperbolic algebras of ranks 9 and 10 have been found, if at all, as subalgebras of $\mathcal{G}_2$. Let us now compare the bounds as obtained by identification as subalgebra with the global bounds of theorem 1.4b, given for rank 7-10. Note that we use the values for roots not in NL^* in the case of the subalgebras as these are the smaller values:

r^2	Rk 7	Rk 8	$\mathcal{G}_3$	Rk 9	Rk 10	$\mathcal{G}_2$
0	5	6	6	7	8	8
-2	21	28	27	36	45	52
-4	71	105	104	148	201	256
-6	217	350	351	534	780	1122
-8	603	1057	1080	1738	2718	4352
-10	1574	2975	3107	5240	8730	15640
-12	3880	7883	8424	14824	26226	52224
-14	9153	19900	21762	39809	74556	165087
-16	20755	48160	53976	102223	202180	495872

TABLE 6.4. Comparison of quality of upper bounds, rank 7 to 10

We see that we only obtain minimal improvements in some rare cases by $\mathcal{G}_3$, and worse bounds throughout by $\mathcal{G}_2$. Therefore we will not list any hyperbolic Lie algebras of rank greater or equal to 7 in this chapter, as we do not obtain improved upper bounds. Nevertheless, the analysis of their root lattices remains valuable as it provides the key to answering questions such as to any kind of subalgebra. Similarly, the decomposition of the Leech lattice has been used in a number of problems since it was established (see [**CS88**], chapter 25).

Let us now proceed to give an account of those hyperbolic Lie algebras which are subalgebras of $\mathcal{G}_7$ and $\mathcal{G}_5$ such that the upper bounds obtained from theorem 1.7 are sharp for some roots of small norm and height. We compare these bounds with the exact values for some imaginary roots, ordered by norm and height. The notation is as before (see section 6.2.1.2), and I calculated the exact root multiplicities by a program based on the Peterson recursion formula for root multiplicities (see [**Kac90**], p.210).

Hyperbolic Lie algebras of Rank 5

$H_1^{(5)} = AE_5 \subset \mathcal{G}_7$

N=7

root-coefficients	norm	mult	bound
0, 1, 1, 1, 1	0	3	3
1, 2, 2, 2, 2	-2	9	9
1, 3, 3, 3, 3	-4	22	22
2, 4, 3, 3, 2	-4	22	22
2, 4, 3, 3, 3	-6	51	51
1, 4, 4, 4, 4	-6	51	51
2, 5, 4, 4, 3	-8	108	108
2, 4, 4, 4, 4	-8	108	108
1, 5, 5, 5, 5	-8	108	108
2, 5, 4, 4, 4	-10	221	221
3, 6, 5, 4, 4	-12	429	432

Hyperbolic Lie algebras of Rank 6

$H_1^{(6)} = AE_6 \subset \mathcal{G}_5$

N=5

root-coefficients	norm	mult	bound
0, 1, 1, 1, 1, 1	0	4	4
1, 2, 2, 2, 2, 2	-2	14	14
2, 4, 3, 3, 2, 2	-4	40	40
1, 3, 3, 3, 3, 3	-4	40	40
2, 4, 3, 3, 3, 3	-6	105	105
1, 4, 4, 4, 4, 4	-6	105	105
2, 5, 4, 4, 3, 3	-8	252	256
2, 4, 4, 4, 4, 4	-8	251	256
2, 5, 4, 4, 4, 4	-10	572	590
3, 6, 5, 4, 4, 3	-10	574	590
3, 6, 5, 4, 3, 4	-10	574	590

$H_6^{(6)} = DE_6 \subset \mathcal{G}_5$

N=5

root-coefficients	norm	mult	bound
0, 1, 2, 1, 1, 1	0	4	4
1, 2, 4, 2, 2, 2	-2	14	14
1, 3, 6, 3, 3, 3	-4	40	40
2, 4, 6, 3, 3, 2	-4	40	40
2, 4, 6, 3, 2, 3	-4	40	40
2, 4, 6, 2, 3, 3	-4	40	40
2, 4, 6, 3, 3, 3	-6	105	105
1, 4, 8, 4, 4, 4	-6	105	105
2, 5, 8, 4, 4, 3	-8	252	256
2, 5, 8, 4, 3, 4	-8	252	256
2, 5, 8, 3, 4, 4	-8	252	256

TABLE 6.5. Root multiplicities of some hyperbolic Lie algebras of rank 5 and 6

6.3. Conclusions

This final section tries to put into perspective the results obtained in the previous sections of chapter 6. In section 6.2 we observed that the root multiplicities of $\mathcal{G}_N$ provided good upper bounds for some hyperbolic Lie algebras while for others they did not bear any resemblance to the correct multiplicities.

We recall that in the cases $N = 5, 7, 11, 23$ there existed examples where the upper bounds obtained in this work represented a significant improvement on the global bounds of theorem 1.4b. Let us therefore turn to these cases. This automatically restricts us to simply laced algebras. Still, we did not obtain useful, sharp upper bounds for each hyperbolic subalgebra contained in one of the $\mathcal{G}_N$. If we recall table 6.3 of $N = 11$ given in section 6.2.2.2 we found that only the rank 4 algebra provided useful bounds whereas the bounds for the rank 3 algebra were far off the true values. We recall from section 6.2.3 that the situation is similar for the remaining N.

One necessary condition for useful upper bounds can be found experimentally. If we consider the numerical evidence for those hyperbolic Lie algebras which contain more than one affine subalgebra we will find that, even for roots of the same small negative norm, the root multiplicities vary considerably. Thus we will never be able to provide sharp bounds which depend on norm only. We therefore now concentrate on the following case:

We consider a simply laced hyperbolic Lie algebra $\mathcal{A}$ with a unique affine subalgebra $\mathcal{A}_0$, thus allowing the definition of a level, as in section 6.2.1.2, above. If we consider numerical examples of successful and useless upper bounds we observe another necessary condition for the construction of useful bounds: They must be sharp for at least some norm 0 vectors. Therefore, we will now determine the exact root multiplicity of norm 0 vectors in hyperbolic Lie algebras and then ask when will this number be equal to the multiplicity of those norm 0 vectors in $\mathcal{G}_N$ which are not elements of NL^* (for the notation, see the corollary to theorem 1.7).

We recall from formula (5.9) that for every simple affine Lie algebra there exists a unique norm 0 vector $\delta = \sum_1^n n_i \alpha_i$ where the n_i are positive integers whose greatest common divisor be 1. For the remainder of this section let $\delta(\mathcal{A}_0)$ denote this norm 0 vector of the affine algebra $\mathcal{A}_0$.

We quote three results of [**Kac90**]:

PROPOSITION 6.1 ([**Kac90**], Prop. 5.10c). *Let $\mathcal{A}$ be a Lie algebra of finite, affine, or hyperbolic type. Let Q denote the lattice spanned by the simple roots. Then the set of all imaginary roots is*

$$\{ \alpha \in Q - \{0\} \mid \alpha^2 \leq 0 \}.$$

REMARK. This is, in fact, an equivalent characterization of the three types of Lie algebras among ordinary Kac-Moody algebras. For every other Kac-Moody algebra, there exist vectors of the root lattice of negative norm which are not roots. Note, however, that the series of GKMs $\mathcal{G}_N$ constructed in this work provides examples of further Lie algebras where all imaginary root lattice vectors are roots.

PROPOSITION 6.2 ([**Kac90**], Prop. 5.7). *Let $\mathcal{A}$ be symmetrizable and let it have Cartan matrix C. Let the Weyl group be denoted W. A root α is isotropic (i.e. $\alpha^2 = 0$) if and only if it is W-equivalent to an imaginary root β such that $\mathrm{supp}\,\beta$ is a subdiagram of affine type of the Dynkin diagram of $\mathcal{A}$. Let the corresponding affine subalgebra be called $\mathcal{A}_0$. Then $\beta = k\delta(\mathcal{A}_0)$.*

PROPOSITION 6.3 ([**Kac90**], Cor. 7.4). *Let $\mathcal{A}_0$ be a simply laced affine Lie algebra of n simple roots. Then the multiplicity of every imaginary root of $\mathcal{A}_0$ is $n-1$.*

We rephrase this for the cases of interest to us.

THEOREM 6.1. *Suppose $\mathcal{A}$ is a simply laced hyperbolic Lie algebra of rank n with a unique affine subalgebra $\mathcal{A}_0$. Then the multiplicity of every isotropic $\alpha \in Q - \{0\}$ is $n-2$.*

PROOF. The Dynkin diagram of the affine subalgebra $\mathcal{A}_0$ must be the Dynkin diagram of $\mathcal{A}$ with one simple root removed, hence it will have $n-1$ simple roots. Now consider $\alpha \neq 0$, such that $\alpha^2 = 0$. By proposition 6.1 it is a root of $\mathcal{A}$. By proposition 6.2 it is W-equivalent to some β which corresponds to some affine subalgebra. Now there is only one such subalgebra, $\mathcal{A}_0$, of $n-1$ simple roots. Hence, by proposition 6.3 the multiplicity of β as a root of $\mathcal{A}_0$ is $n-2$. Hence its multiplicity as a root of $\mathcal{A}$ is $n-2$. This in turn implies that the multiplicity of α is $n-2$. □

Theorem 1.1 of [**Bor90a**] shows that, for any root r of $\mathcal{G}_N$ with $r^2 > 0$, there exists a unique representation as a sum of positive simple roots. As we will see below, no such uniqueness holds for roots r such that $r^2 \leq 0$. Instead, we have the following situation. Let $\mathcal{A}$ be a hyperbolic Lie algebra with unique affine subalgebra $\mathcal{A}_0$ such that $\mathcal{A}$ is subalgebra of one of the GKMs $\mathcal{G}_N$. As before, let $\mathcal{R} = \{r_i\}$ denote the set of real simple roots of $\mathcal{G}_N$. The vector $\delta = \delta(\mathcal{A}_0)$ can be represented as $\delta = \sum_j n_{i_j} r_{i_j}$ where the r_{i_j}, $j = 1, \ldots, n$ are the simple roots of $\mathcal{A}_0$. The vector δ lies within the (closure of the) Weyl chamber of the affine Lie algebra $\mathcal{A}_0$. It also lies within the (closure of the) Weyl chamber of the GKM $\mathcal{G}_N$. Suppose, it did not. Then there would exist a root r of $\mathcal{G}_N$ such that the reflection ϕ_r takes δ to a root of smaller height. As δ lies within the Weyl chamber of $\mathcal{A}_0$ r must be distinct from any of the simple roots r_{i_j}, $j = 1, \ldots, n$. Then $\phi_r(\delta)$ is a root which is neither positive nor negative, a contradiction. We conclude that if $\delta = \sum_j n_{i_j} r_{i_j} = \sum_k n_{i_k} r_{i_k}$ then by proposition 6.2 not only the collection of r_{i_j} but also the collection of r_{i_k} corresponds to an affine subalgebra of $\mathcal{G}_N$.

The multiplicity of δ as a root of $\mathcal{G}_N$ and the multiplicity of δ as a root of the hyperbolic Lie algebra will be equal if and only if the representation $\delta = \sum n_{i_j} r_{i_j}$ is unique in $\mathcal{G}_N$. We recall formula (5.13) for the generalized centre c of an affine hole and formula (5.10) for the relation $\delta^\vee = \frac{1}{d}\nu^{-1}(\delta)$, using all notation as in chapter 5. Then

$$(c, 1, *) = \frac{\delta^\vee}{h^\vee} = \frac{\nu^{-1}(\delta)}{dh^\vee}.$$

Thus we can identify the centre c from δ alone, using the fact that $(c, 1, *)$ must have height 1. This can be reformulated as follows: If $\delta = \sum_j n_{i_j} r_{i_j} = \sum_k n_{i_k} r_{i_k}$ within the GKM $\mathcal{G}_N$, then the representatives of r_{i_j} and those of r_{i_k} are associated to the same affine hole of $\mathcal{R}$. To confirm that δ can only be represented in a unique way it therefore suffices to check the the products $dh^\vee$ of the components of the affine hole in $\mathcal{R}$ determined by δ.

THEOREM 6.2. *Suppose that a simply laced hyperbolic Lie algebra $\mathcal{A}$ with unique affine subalgebra $\mathcal{A}_0$ is contained in one of the GKMs $\mathcal{G}_N$ constructed in* theorem 1.6. *Consider the affine hole H of $\mathcal{R}$ containing the elements of $\mathcal{R}$ representing $\delta(\mathcal{A}_0)$. If the product $d(\mathcal{A}_0)h^\vee(\mathcal{A}_0)$ is smaller than the product $dh^\vee$ of any other component of H then the upper bound for the multiplicity of δ in $\mathcal{A}$ is sharp.*

□

REMARK. By theorem 6.1, the multiplicity of any norm 0 vector in $\mathcal{A}$ equals that of δ. Suppose we obtain a sharp upper bound for the multiplicity of δ. If the upper bounds depend on norm only they will be sharp for every norm 0 vector. However, in the cases $\mathcal{G}_N$ at hand, we obtain discrepancies for norm 0 vectors which are elements of NL^*.

EXAMPLES. It is now straightforward to understand the quality of the upper bounds obtained in section 6.2. For example, A_1 in $A_1\ \mathbf{23}A_1$ satisfies the conditions of theorem 6.2, A_1 in $A_1^2\ \mathbf{11}A_1^2$ does not. Correspondingly, the bounds for AE_3 as a subset of $\mathcal{G}_{23}$ are useful, those for AE_3 as a subset of $\mathcal{G}_{11}$ are not.

Similarly, A_2 in $A_2\ \mathbf{11}A_2$ satisfies the condition. Hence the bounds for AE_4 obtained from $N = 11$ are useful. As a last example, the bounds for $H_{71}^{(3)}$ are useless, again because $A_1^2\ \mathbf{11}A_1^2$ does not satisfy the conditions of the theorem.

Let us look at the results of this work from the point of view that it provides a strategy to calculate sharp upper bounds for the root multiplicities of some hyperbolic Lie algebras. We are now in the position to conjecture how far it may be possible to generalize this strategy. Suppose we consider a hyperbolic Lie algebra $\mathcal{A}$ of small rank, that is, the global upper bounds as provided by theorem 1.4b are substantially greater than the true values. We will have to find a suitable automorphism σ of the Leech lattice such that the root system $\mathcal{R}$ contains $\mathcal{A}$ as a subalgebra and such that at the same time the conditions of theorem 6.2 are satisfied. Recall that we can search for such σ easily as we understand the action of σ on Dynkin diagrams directly (chapter 5.4). Note that theorem 6.2 provides some idea of the quality of the bounds before we begin to calculate denominator formulas.

The twisted denominator formula will then describe a Lie superalgebra, as indicated in the final remark of section 1.6. We will be able to obtain upper bounds for the root multiplicities of $\mathcal{A}$ if the twisted denominator formula describes in fact either a GKM or a Lie superalgebra with strictly alternating multiplicities. Borcherds states some conjectures about the properties of such automorphisms in chapter 6 of [**Bor90b**].

For all hyperbolic Lie algebras of rank 7 to 10, the root multiplicities for some roots of small negative norm are collected in appendix B, and compared with both theorem 1.4b and the results of this work. One remarkable result of the calculations concerns the algebra $T_{4,3,3}$ which contains roots of multiplicities both greater and smaller than p_6, disproving a number of conjectures. The data shows furthermore that for simply laced hyperbolic Lie algebras of high rank the upper bound of theorem 1.4b already proves useful, which is one of the reasons why the new upper bounds could not improve on them. However, in the cases of those BE_n and CE_n which we identified as subalgebras of the $\mathcal{G}_N$, the new bounds are substantially greater than the true values, and theorem 1.4b does not apply. We note that the bounds for CE_n are not even sharp for the some of the norm 0 vectors. A look on the numerical results for the BE_n shows that, even though the bounds for norm 0 are sharp, those for any other norm are useless. This provides a note of caution: Constructing sharp bounds for norm 0 vectors is necessary if we want to obtain useful upper bounds but not sufficient. The question of useful upper bounds remains wide open in these cases.

Acknowledgements: I wish to thank Richard Borcherds, my research supervisor, for suggesting many of the problems discussed in this work and for the advice and encouragement he provided throughout my research. Furthermore, I would like to thank Simon Norton for his patient explanations of the ATLAS, and Elizabeth Jurisich for clarifications regarding the nature of specializations. Finally, I would like to thank the referee for valuable comments on the first version of this paper.

Appendix A

This appendix contains the complete tables of affine and finite diagrams for the sets $\mathcal{R}$ as described in chapter 5. The conventions denoting Lie algebras are standard with capital letters denoting affine Lie algebras and small letters denoting finite ones. The presence of a bold-faced integer $\mathbf{N}$ in front of a component indicates that this component consists of long roots (the ratio of the norms of being N to 1).

Each table lists first all affine and then all finite diagrams. The types of diagrams are in the first instance ordered by their indices. They are ordered by natural order of partitions. This convention is adopted even in those cases (like $\Delta^{(2)}$ or $\Delta^{(3)}$) where the index does not coincide with the rank. For a fixed set of indices, the order is alphabetic in the components. Note that the alphabet is extended such that any component $\mathbf{N}\Delta$ immediately follows Δ.

The second column of each table contains the size of the automorphism group of the respective diagrams. The total number of representatives of any diagram within a fundamental volume of the respective fixed point lattice is the size of the automorphism group of the lattice divided by the automorphism group of the diagram. The third column contains the unit volume of any one representative of a diagram. The necessary calculations are described in chapter 5.3. The total volume is the product of the unit volume by the total number of representatives of a diagram. The volume formula (theorem 5.6) states that the total volumes of the complete list of types must add up to the volume of the fundamental region. This equals $\sqrt{N}^M$ as follows from the explicit bases provided in chapter 6.2. Here, $N = 23, 11, 7, 5, 3, 2$, and $M = 24/(N+1)$. The unit volume and total volume are given in units indicated at the top of each table. These units are chosen purely for convenience.

N=23

Order of the automorphism group of the fixed point lattice: 2.
There are 1 type of affine diagram and 2 types of finite diagrams.

type of diagram	group	unit volume $[u=2!^{-1}\sqrt{23}^{-1}]$	total volume $[u=\sqrt{23}^{-1}]$
A_1 $\mathbf{23}A_1$	2	8	4
a_2 $\mathbf{23}a_1$	1	9	9
a_1^2 $\mathbf{23}a_1$	1	10	10
			23

Total volume of 3 types (units of 1) = $23/\sqrt{23} = \sqrt{23}^1$.

N=11

Order of the automorphism group of the fixed point lattice: $2^3\, 3^1 = 24$.
There are 2 types of affine diagrams and 6 types of finite diagrams.

type of diagram	group	unit volume $[u=(11*4!)^{-1}]$	total volume $[u=11^{-1}]$
$A_2\ \ 11A_2$	6	27	4.5
$A_1{}^2\ \ 11A_1{}^2$	8	64	8
$a_3\ \ 11a_2$	1	18	18
$a_3\ \ a_1\ \ 11a_1$	1	44	44
$a_3\ \ 11a_1{}^2$	2	16	8
$a_2\ \ 11a_2\ \ a_1$	2	21	10.5
$a_2\ \ a_1\ \ 11a_1{}^2$	1	18	18
$a_1{}^3\ \ 11a_1{}^2$	2	20	10
			121

Total volume of 8 types (units of 1) = $121/11 = \sqrt{11}^2$.

N=7

Order of the automorphism group of the fixed point lattice: $2^5\, 3^1\, 7^1 = 672$.
There 3 types of affine diagrams and 11 types of finite diagrams.

type of diagram	group	unit volume $[u=(7^2*6!)^{-1}\sqrt{7}]$	total volume $[u=\frac{1}{315}\sqrt{7}]$
A_6	14	343	147
$A_3\ \ 7A_3$	8	64	48
$A_1{}^3\ \ 7A_1{}^3$	48	512	64
$a_6\ \ 7a_1$	2	147	441
$a_5\ \ 7a_2$	2	63	189
$a_4\ \ 7a_3$	1	30	180
$d_4\ \ 7a_3$	2	28	84
$a_4\ \ a_1\ \ 7a_1{}^2$	1	70	420
$d_4\ \ 7a_1{}^3$	6	28	28
$a_3\ \ 7a_3\ \ a_1$	2	36	108
$a_3\ \ a_1\ \ 7a_1{}^3$	2	32	96
$a_2\ \ a_1{}^3\ \ 7a_1{}^2$	2	84	252
$a_2\ \ a_1{}^2\ \ 7a_1{}^3$	2	36	108
$a_1{}^4\ \ 7a_1{}^3$	6	40	40
			2205

Total volume of 14 types (units of 1) = $2205\frac{\sqrt{7}}{315} = \sqrt{7}^3$.

N=5

Order of the automorphism group of the fixed point lattice: $2^7\, 3^2\, 5^2 = 28800$.
There are 6 types of affine diagrams and 24 types of finite diagrams.

type of diagram	group	unit volume $[u=(25*8!)^{-1}]$	total volume $[u=1/210]$
D_6 $5A_1{}^2$	8	800	600
$A_4{}^2$	200	3125	93.75
A_4 $5A_4$	20	125	37.5
D_4 $5D_4$	24	144	36
$A_2{}^2$ $5A_2{}^2$	72	729	60.75
$A_1{}^4$ $5A_1{}^4$	384	4096	64
d_6 $5a_3$	2	80	240
d_6 $5a_2$ $5a_1$	1	90	540
d_6 $5a_1{}^3$	2	100	300
a_5 $5a_4$	2	45	135
d_5 $5a_4$	2	40	120
d_5 $5d_4$	2	40	120
a_5 $5a_2{}^2$	2	45	135
$a_4{}^2$ $5a_1$	8	375	281.25
a_4 $5a_4$ a_1	4	55	82.5
d_4 $5d_4$ a_1	6	52	52
a_4 a_3 $5a_1{}^2$	4	200	300
a_4 $5a_3$ $a_1{}^2$	4	120	180
a_4 a_2 $5a_2$ a_1	4	225	337.5
a_4 a_2 $5a_2$ $5a_1$	4	105	157.5
$5d_4$ a_2 $a_1{}^3$	6	84	84
d_4 $a_1{}^2$ $5a_1{}^3$	6	120	120
d_4 a_1 $5a_1{}^4$	6	56	56
a_3 a_2 $5a_2{}^2$	2	54	162
a_3 a_2 $5a_2$ a_1 $5a_1$	2	120	360
a_3 a_2 a_1 $5a_1{}^3$	2	120	360
a_3 $a_1{}^2$ $5a_1{}^4$	4	64	96
$a_2{}^2$ $5a_2{}^2$ a_1	8	63	47.25
a_2 $a_1{}^3$ $5a_1{}^4$	6	72	72
$a_1{}^5$ $5a_1{}^4$	24	80	20
			5250

Total volume of 30 types (units of 1) = $5250/210 = 25 = \sqrt{5}^4$.

N=3

Order of the automorphism group of the fixed point lattice: $2^{10}\, 3^7\, 5^1\, 7^1 = 78382080$. There are 15 types of affine diagrams and 93 types of finite diagrams.

type of diagram			group	unit volume $[u=(27*12!)^{-1}]$	total volume $[u=1/118800]$
A_8	$\mathbf{3A_2}^2$		36	2187	43740
E_7	$\mathbf{3A_5}$		4	648	116640
D_7	$\mathbf{3A_3}$	G_2	8	2304	207360
A_6	$\mathbf{3A_6}$		14	343	17640
D_6	$\mathbf{3D_6}$		8	400	36000
E_6	$\mathbf{3E_6}$		12	432	25920
E_6	G_2^3		36	6912	138240
A_5	$\mathbf{3A_5}$	$D_4^{(3)}$	24	1296	38880
A_5	D_4	$\mathbf{3A_1}^3$	288	20736	51840
$D_4^{(3)6}$			720	46656	46656
D_4	$\mathbf{3D_4}$	$D_4^{(3)2}$	48	5184	77760
A_3^2	$\mathbf{3A_3}^2$		128	4096	23040
A_2^6			349920	531441	1093.5
A_2^3	$\mathbf{3A_2}^3$		1944	19683	7290
A_1^6	$\mathbf{3A_1}^6$		46080	262144	4096
a_8	$\mathbf{3a_5}$		2	135	48600
a_8	$\mathbf{3a_5}$		2	135	48600
a_8	$\mathbf{3a_3}$	$\mathbf{3a_2}$	1	162	116640
a_8	$\mathbf{3a_2}^2$	$\mathbf{3a_1}$	4	189	34020
a_7	$\mathbf{3a_6}$		1	84	60480
d_7	$\mathbf{3a_6}$		1	70	50400
d_7	$\mathbf{3d_6}$		2	64	23040
e_7	$\mathbf{3a_6}$		1	63	45360
e_7	$\mathbf{3d_6}$		1	54	38880
e_7	$\mathbf{3e_6}$		2	51	18360
a_7	$\mathbf{3a_5}$	$\mathbf{3a_1}$	2	96	34560
d_7	$\mathbf{3a_5}$	a_1	1	144	103680
d_7	$\mathbf{3d_5}$	a_1	2	132	47520
e_7	$\mathbf{3a_5}$	$\mathbf{3a_1}$	2	78	28080
d_7	$\mathbf{3a_4}$	g_2	1	90	64800
d_7	$\mathbf{3d_4}$	g_2	2	84	30240
a_7	$\mathbf{3a_3}^2$		2	96	34560
d_7	$\mathbf{3a_3}^2$		2	88	31680
d_7	$\mathbf{3a_3}$	$\mathbf{3a_2}\ a_1$	2	180	64800
d_7	$\mathbf{3a_3}$	$g_2\ \mathbf{3a_1}$	2	108	38880
a_6	$\mathbf{3a_6}$	a_1	2	105	37800
a_6	$\mathbf{3e_6}$	a_1	2	105	37800
d_6	$\mathbf{3d_6}$	a_1	2	84	30240
e_6	$\mathbf{3e_6}$	a_1	4	75	13500
a_6	$\mathbf{3a_5}$	g_2	2	63	22680

type of diagram	group	unit volume $[u=(27*12!)^{-1}]$	total volume $[u=1/118800]$
d_6 $\mathbf{3}a_5$ g_2	2	54	19440
e_6 $\mathbf{3}a_5$ g_2	4	51	9180
$\mathbf{3}e_6$ a_5 a_2	4	117	21060
$\mathbf{3}d_6$ d_5 $a_1{}^2$	2	112	40320
e_6 $\mathbf{3}d_5$ $a_1{}^2$	4	156	28080
$\mathbf{3}d_6$ a_4 a_3	2	120	43200
d_6 $\mathbf{3}d_4$ g_2 $\mathbf{3}a_1$	2	60	21600
$\mathbf{3}d_6$ d_4 a_2 a_1	2	132	47520
e_6 $\mathbf{3}a_4$ g_2 a_1	2	105	37800
e_6 $\mathbf{3}d_4$ $a_1{}^3$	12	324	19440
e_6 $\mathbf{3}a_3$ $g_2{}^2$	4	66	11880
$\mathbf{3}e_6$ a_3 $a_2{}^2$	4	162	29160
e_6 $\mathbf{3}a_3$ g_2 $a_1{}^2$	4	216	38880
e_6 $\mathbf{3}a_2$ $g_2{}^2$ a_1	4	135	24300
e_6 $g_2{}^3$ $\mathbf{3}a_1$	12	81	4860
a_5 $\mathbf{3}a_5$ a_3	4	180	32400
a_5 $\mathbf{3}d_5$ a_3	4	180	32400
a_5 $\mathbf{3}a_5$ a_2 $\mathbf{3}a_1$	4	126	22680
a_5 $\mathbf{3}a_5$ g_2 a_1	4	78	14040
a_5 d_4 $\mathbf{3}d_4$	12	252	15120
a_5 d_4 $\mathbf{3}a_3$ $\mathbf{3}a_1$	4	288	51840
d_5 $\mathbf{3}d_4$ $g_2{}^2$	4	40	7200
$\mathbf{3}d_5$ d_4 $a_2{}^2$	4	216	38880
a_5 d_4 $\mathbf{3}a_2$ $\mathbf{3}a_1{}^2$	4	324	58320
a_5 d_4 $\mathbf{3}a_1{}^4$	12	360	21600
d_5 $\mathbf{3}d_4$ $a_1{}^3$ $\mathbf{3}a_1$	6	240	28800
a_5 a_3 $\mathbf{3}a_3$ $\mathbf{3}a_1{}^2$	4	216	38880
a_5 a_3 $\mathbf{3}a_2{}^2$ $\mathbf{3}a_1$	4	216	38880
d_5 $\mathbf{3}a_3$ $g_2{}^2$ a_1	4	84	15120
a_5 $\mathbf{3}a_3$ g_2 $a_1{}^2$ $\mathbf{3}a_1$	4	168	30240
a_5 a_2 $\mathbf{3}a_2{}^3$	12	135	8100
a_5 a_2 $\mathbf{3}a_2{}^3$	12	135	8100
a_5 $g_2{}^4$	48	63	945
d_5 $g_2{}^3$ $\mathbf{3}a_1{}^2$	12	60	3600
a_5 g_2 $a_1{}^3$ $\mathbf{3}a_1{}^3$	12	360	21600
a_4 d_4 $\mathbf{3}d_4$ $\mathbf{3}a_1$	6	180	21600
a_4 d_4 $\mathbf{3}a_3$ $\mathbf{3}a_2$	2	180	64800
d_4 $\mathbf{3}d_4$ a_3 g_2	6	120	14400
d_4 $\mathbf{3}d_4$ a_3 $\mathbf{3}a_1{}^2$	12	128	7680
d_4 $\mathbf{3}d_4$ a_2 g_2 $\mathbf{3}a_1$	6	84	10080
d_4 $\mathbf{3}d_4$ $g_2{}^2$ a_1	12	52	3120
a_4 $\mathbf{3}d_4$ g_2 $a_1{}^3$	6	180	21600
d_4 $\mathbf{3}d_4$ $a_1{}^3$ $\mathbf{3}a_1{}^2$	12	176	10560
a_4 a_3 $\mathbf{3}a_3{}^2$	2	120	43200
d_4 a_3 $\mathbf{3}a_3{}^2$	4	112	20160

APPENDIX A

type of diagram	group	unit volume $[u=(27*12!)^{-1}]$	total volume $[u=1/118800]$
$d_4\ a_3\ \mathbf{3}a_3\ a_1\ \mathbf{3}a_1^2$	4	240	43200
$\mathbf{3}d_4\ a_3\ g_2\ a_1^3\ \mathbf{3}a_1$	6	128	15360
$d_4\ a_3\ a_1\ \mathbf{3}a_1^5$	12	288	17280
$a_4\ g_2^4\ \mathbf{3}a_1$	24	45	1350
$d_4\ g_2^3\ \mathbf{3}a_1^3$	36	44	880
$\mathbf{3}d_4\ a_2\ g_2^2\ a_1^3$	12	84	5040
$d_4\ a_1^3\ \mathbf{3}a_1^6$	36	224	4480
$a_3^2\ \mathbf{3}a_3^2\ a_1$	8	144	12960
$a_3\ a_2^2\ \mathbf{3}a_2^3$	12	162	9720
$a_3\ g_2^5$	120	30	180
$a_3\ g_2^4\ \mathbf{3}a_1^2$	48	32	480
$\mathbf{3}a_3\ a_2^2\ g_2^2\ a_1^2$	8	144	12960
$a_3\ g_2\ a_1^4\ \mathbf{3}a_1^4$	24	288	8640
$a_3\ a_1^4\ \mathbf{3}a_1^6$	48	256	3840
$a_2^6\ \mathbf{3}a_1$	1440	2187	1093.5
$a_2^4\ g_2^2\ \mathbf{3}a_1$	48	405	6075
$a_2^3\ \mathbf{3}a_2^3\ a_1$	72	189	1890
$a_2^3\ \mathbf{3}a_2\ g_2^2\ a_1$	12	243	14580
$a_2\ g_2^5\ \mathbf{3}a_1$	120	21	126
$g_2^6\ a_1$	720	13	13
$a_2\ a_1^5\ \mathbf{3}a_1^6$	120	288	1728
$g_2\ a_1^6\ \mathbf{3}a_1^5$	120	224	1344
$a_1^7\ \mathbf{3}a_1^6$	720	320	320
			3207600

Total volume of 108 types (units of 1) = $3207600/118800 = 27 = \sqrt{3}^6$.

N=2

Order of the automorphism group of the fixed point lattice: $2^{21} \ 3^5 \ 5^2 \ 7^1 = 89181388800$.

There are 40 types of affine diagrams and 435 types of finite diagrams.

type of diagram	group	unit volume $[u=(16*16!)^{-1}]$	total volume $[u=1891890^{-1}]$
$A_{17}^{(2)} \ \mathbf{2}E_7$	2	648	163296
$A_{15}^{(2)} \ D_9^{(2)}$	2	512	129024
$A_{11}^{(2)} \ D_7^{(2)} \ E_6^{(2)}$	2	3456	870912
$D_{10}^{(2)} \ E_7$	2	648	163296
$C_{10} \ B_6$	2	484	121968
$A_9^{(2)\,2} \ \mathbf{2}D_6$	8	4000	252000
$D_9 \ \mathbf{2}A_7$	8	1024	64512
$A_9 \ \mathbf{2}A_4 \ B_3$	20	5000	126000
$A_8 \ \mathbf{2}A_8$	18	729	20412
$B_8 \ E_8$	1	900	453600
$D_8 \ \mathbf{2}D_8$	4	784	98784
$E_8 \ \mathbf{2}E_8$	1	900	453600
$C_8 \ F_4^{\,2}$	4	2916	367416
$D_8 \ B_4^{\,2}$	8	5488	345744
$\mathbf{2}A_7 \ A_7^{(2)} \ D_5$	16	4096	129024
$A_7^{(2)\,2} \ D_5^{(2)\,2}$	16	16384	516096
$A_7 \ \mathbf{2}D_5 \ D_5^{(2)}$	16	4096	129024
$E_7 \ B_5 \ F_4$	2	5832	1469664
$A_7 \ \mathbf{2}A_3 \ C_3^{\,2}$	64	16384	129024
$E_6 \ \mathbf{2}E_6 \ E_6^{(2)}$	6	5184	435456
$E_6^{(2)\,4}$	24	20736	435456
$D_6 \ D_6^{(2)\,2}$	8	4000	252000
$C_6^{\,2} \ B_4$	8	2744	172872
$E_6 \ \mathbf{2}A_5 \ C_5$	12	5184	217728
$D_6 \ C_4 \ B_3^{\,2}$	16	20000	630000
$A_5 \ \mathbf{2}A_5 \ A_5^{(2)} \ D_4^{(2)}$	24	15552	326592
$A_5^{(2)\,4} \ \mathbf{2}D_4$	192	62208	163296
$D_5^{\,2} \ \mathbf{2}A_3^{\,2}$	128	32768	129024
$A_5^{\,2} \ \mathbf{2}A_2^{\,2} \ C_2$	288	69984	122472
$D_4 \ D_4^{(2)\,4}$	192	62208	163296
$A_4^{\,2} \ \mathbf{2}A_4^{\,2}$	200	15625	39375
$C_4^{\,4}$	192	10000	26250
$D_4^{\,2} \ \mathbf{2}D_4^{\,2}$	384	20736	27216
$D_4^{\,2} \ C_2^{\,4}$	1536	186624	61236
$D_3^{(2)\,8}$	21504	1048576	24576
$A_3^{\,2} \ \mathbf{2}A_3^{\,2} \ D_3^{(2)\,2}$	512	131072	129024

APPENDIX A

type of diagram	group	unit volume $[u=(16*16!)^{-1}]$	total volume $[u=1891890^{-1}]$
$A_3^4\ \mathbf{2}A_1^4$	24576	1048576	21504
$A_2^4\ \mathbf{2}A_2^4$	15552	531441	17222.625
A_1^{16}	660602880	268435456	204.8
$A_1^8\ \mathbf{2}A_1^8$	2752512	16777216	3072
$c_{10}\ b_7$	1	70	35280
$c_{10}\ \mathbf{2}e_7$	1	56	28224
$c_{10}\ b_6\ \mathbf{2}a_1$	1	92	46368
$c_{10}\ \mathbf{2}e_6\ a_1$	1	114	57456
$c_{10}\ \mathbf{2}d_5\ b_2$	1	120	60480
$c_{10}\ \mathbf{2}a_4\ b_3$	1	130	65520
$c_{10}\ b_4\ \mathbf{2}a_2\ \mathbf{2}a_1$	1	144	72576
$a_9\ \mathbf{2}a_8$	1	135	68040
$a_9\ b_8$	1	130	65520
$a_9\ \mathbf{2}e_8$	1	115	57960
$b_9\ a_8$	1	99	49896
$b_9\ e_8$	1	47	23688
$c_9\ b_8$	1	50	25200
$d_9\ \mathbf{2}a_8$	1	108	54432
$d_9\ \mathbf{2}d_8$	1	88	44352
$d_9\ \mathbf{2}e_8$	1	76	38304
$b_9\ d_7\ a_1$	1	96	48384
$b_9\ e_7\ a_1$	1	74	37296
$c_9\ \mathbf{2}e_7\ a_1$	1	74	37296
$d_9\ \mathbf{2}a_7\ \mathbf{2}a_1$	2	136	34272
$b_9\ a_6\ a_2$	1	147	74088
$b_9\ d_6\ a_2$	1	114	57456
$c_9\ b_6\ \mathbf{2}a_2$	1	72	36288
$c_9\ \mathbf{2}d_6\ b_2$	1	80	40320
$a_9\ \mathbf{2}a_6\ a_1\ \mathbf{2}a_1$	1	280	141120
$a_9\ \mathbf{2}e_6\ a_1\ \mathbf{2}a_1$	1	240	120960
$a_9\ \mathbf{2}a_5\ b_3$	1	180	90720
$a_9\ \mathbf{2}d_5\ b_3$	1	160	80640
$b_9\ a_5\ a_3$	1	156	78624
$c_9\ \mathbf{2}a_5\ b_3$	1	90	45360
$a_9\ \mathbf{2}a_4^2$	2	175	44100
$a_9\ \mathbf{2}a_4\ b_4$	2	170	42840
$a_9\ \mathbf{2}a_4\ b_3\ \mathbf{2}a_1$	2	220	55440
$b_9\ a_4\ a_3\ a_1$	1	200	100800
$c_9\ b_4\ \mathbf{2}a_3\ \mathbf{2}a_1$	1	104	52416
$a_9\ \mathbf{2}a_4\ \mathbf{2}a_2\ a_1\ \mathbf{2}a_1$	2	360	90720
$a_8\ \mathbf{2}a_8\ a_1$	2	171	43092
$a_8\ \mathbf{2}e_8\ a_1$	1	153	77112
$b_8\ c_8\ a_1$	1	66	33264
$b_8\ d_8\ a_1$	1	124	62496
$b_8\ e_8\ \mathbf{2}a_1$	1	62	31248

$N=2$

type of diagram	group	unit volume $[u=(16*16!)^{-1}]$	total volume $[u=1891890^{-1}]$
d_8 $2d_8$ a_1	1	116	58464
e_8 $2e_8$ a_1	1	61	30744
b_8 a_7 a_2	1	192	96768
b_8 e_7 a_2	1	114	57456
b_8 e_7 b_2	1	66	33264
c_8 b_7 a_2	1	102	51408
e_8 $2e_7$ b_2	1	64	32256
$2e_8$ e_7 a_2	1	93	46872
c_8 $2a_7$ a_1^2	2	136	34272
d_8 $2a_7$ a_1^2	2	240	60480
$2d_8$ e_7 a_1^2	1	124	62496
b_8 d_6 a_3	1	160	80640
b_8 e_6 c_3	1	72	36288
c_8 b_6 b_3	1	54	27216
c_8 b_6 c_3	1	76	38304
$2d_8$ e_6 a_3	1	132	66528
e_8 $2e_6$ b_3	1	69	34776
$2e_8$ e_6 a_3	1	126	63504
a_8 $2e_6$ $2a_2$ a_1	1	189	95256
b_8 c_6 a_2 a_1	1	102	51408
c_8 $2a_6$ a_2 a_1	1	210	105840
d_8 b_6 b_2 $2a_1$	1	136	68544
d_8 $2d_6$ b_2 $2a_1$	1	128	64512
$2d_8$ d_6 a_2 a_1	1	180	90720
e_8 b_6 $2a_2$ $2a_1$	1	96	48384
$2e_8$ a_6 a_2 a_1	1	231	116424
a_8 $2d_5$ b_4	1	108	54432
b_8 a_5 a_4	1	210	105840
b_8 c_5 a_4	1	90	45360
b_8 d_5 c_4	1	80	40320
c_8 b_5 f_4	1	54	27216
d_8 b_5 b_4	1	92	46368
d_8 $2d_5$ b_4	1	88	44352
$2d_8$ d_5 a_4	1	160	80640
$2d_8$ d_5 d_4	1	144	72576
e_8 b_5 $2a_4$	1	85	42840
e_8 $2d_5$ b_4	1	76	38304
$2e_8$ a_5 a_4	1	195	98280
$2e_8$ d_5 a_4	1	160	80640
b_8 a_5 a_3 a_1	1	264	133056
c_8 $2a_5$ c_3 a_1	1	156	78624
c_8 $2a_5$ a_2^2	2	324	81648
d_8 $2a_5$ b_2 a_1 $2a_1$	1	264	133056
c_8 $2a_4$ f_4 a_1	1	110	55440
c_8 b_4^2 $2a_1$	2	68	17136

APPENDIX A

type of diagram	group	unit volume $[u=(16*16!)^{-1}]$	total volume $[u=1891890^{-1}]$
$c_8\ f_4^{\,2}\ \mathbf{2}a_1$	2	76	19152
$d_8\ \mathbf{2}a_4\ b_4\ a_1$	1	180	90720
$d_8\ b_4^{\,2}\ \mathbf{2}a_1$	2	120	30240
$c_8\ \mathbf{2}a_4\ c_3\ a_2$	1	240	120960
$c_8\ f_4\ \mathbf{2}a_3\ a_2$	1	168	84672
$c_8\ f_4\ c_3\ \mathbf{2}a_2$	1	120	60480
$d_8\ b_4\ \mathbf{2}a_2\ b_2\ \mathbf{2}a_1$	1	192	96768
$c_8\ \mathbf{2}a_3\ c_3^{\,2}$	2	176	44352
$d_8\ \mathbf{2}a_3\ b_2^{\,2}\ \mathbf{2}a_1^{\,2}$	2	288	72576
$b_7\ e_7\ c_3$	1	84	42336
$e_7\ \mathbf{2}e_7\ c_3$	1	70	35280
$\mathbf{2}a_7\ d_7\ b_2\ a_1$	2	152	38304
$\mathbf{2}a_7\ e_7\ b_2\ a_1$	1	136	68544
$c_7\ \mathbf{2}e_7\ a_2\ a_1$	1	114	57456
$\mathbf{2}d_7\ e_7\ a_2\ a_1$	1	192	96768
$a_7\ \mathbf{2}a_7\ a_1^{\,2}\ \mathbf{2}a_1$	2	256	64512
$\mathbf{2}a_7\ d_6\ c_4$	2	104	26208
$\mathbf{2}a_7\ e_6\ c_4$	1	96	48384
$b_7\ d_6\ c_4$	1	100	50400
$c_7\ b_6\ f_4$	1	38	19152
$\mathbf{2}d_7\ e_6\ a_4$	1	180	90720
$e_7\ b_6\ f_4$	1	58	29232
$e_7\ \mathbf{2}e_6\ f_4$	1	51	25704
$\mathbf{2}e_7\ c_6\ a_4$	1	100	50400
$\mathbf{2}e_7\ e_6\ c_4$	1	78	39312
$a_7\ b_6\ c_3\ \mathbf{2}a_1$	1	160	80640
$a_7\ \mathbf{2}e_6\ b_3\ a_1$	1	144	72576
$\mathbf{2}a_7\ c_6\ a_3\ a_1$	2	152	38304
$\mathbf{2}a_7\ d_6\ a_3\ a_1$	2	272	68544
$b_7\ c_6\ a_3\ a_1$	1	144	72576
$c_7\ \mathbf{2}d_6\ c_3\ a_1$	1	92	46368
$e_7\ b_6\ b_3\ \mathbf{2}a_1$	1	76	38304
$e_7\ \mathbf{2}d_6\ c_3\ a_1$	1	144	72576
$a_7\ \mathbf{2}d_6\ b_2^{\,2}$	2	144	36288
$c_7\ \mathbf{2}e_6\ a_2^{\,2}$	2	180	45360
$d_7\ b_6\ \mathbf{2}a_2\ b_2$	1	108	54432
$e_7\ \mathbf{2}a_6\ a_2\ b_2$	1	210	105840
$\mathbf{2}a_7\ c_5\ d_5$	2	104	26208
$\mathbf{2}a_7\ d_5^{\,2}$	2	192	48384
$b_7\ a_5\ c_5$	1	120	60480
$d_7\ b_5^{\,2}$	2	64	16128
$\mathbf{2}d_7\ d_5^{\,2}$	2	176	44352
$e_7\ b_5^{\,2}$	1	54	27216
$\mathbf{2}e_7\ c_5\ d_5$	1	88	44352
$a_7\ \mathbf{2}d_5\ b_4\ a_1$	2	136	34272

type of diagram	group	unit volume $[u=(16*16!)^{-1}]$	total volume $[u=1891890^{-1}]$
a_7 $\mathbf{2}d_5$ f_4 a_1	2	152	38304
$\mathbf{2}a_7$ d_5 a_4 $\mathbf{2}a_1$	2	200	50400
$\mathbf{2}a_7$ d_5 c_4 a_1	2	136	34272
c_7 $\mathbf{2}a_5$ f_4 a_1	1	78	39312
e_7 b_5 $\mathbf{2}a_4$ a_1	1	110	55440
e_7 b_5 f_4 $\mathbf{2}a_1$	1	76	38304
e_7 $\mathbf{2}d_5$ f_4 a_1	1	104	52416
a_7 $\mathbf{2}a_5$ c_3 b_2	1	192	96768
a_7 $\mathbf{2}d_5$ b_3 a_2	2	216	54432
a_7 $\mathbf{2}d_5$ c_3 b_2	2	176	44352
c_7 $\mathbf{2}d_5$ c_3 a_2	1	144	72576
e_7 $\mathbf{2}a_5$ c_3 b_2	1	156	78624
e_7 b_5 $\mathbf{2}a_3$ a_2	1	168	84672
e_7 b_5 c_3 $\mathbf{2}a_2$	1	120	60480
a_7 $\mathbf{2}a_5$ b_3 a_1 $\mathbf{2}a_1$	1	192	96768
$\mathbf{2}a_7$ a_5 a_3 a_1 $\mathbf{2}a_1$	2	288	72576
e_7 $\mathbf{2}a_5$ b_3 a_1 $\mathbf{2}a_1$	1	156	78624
$\mathbf{2}a_7$ d_5 a_2 b_2 a_1	2	216	54432
$\mathbf{2}a_7$ a_5 b_2 a_1^3	2	288	72576
a_7 $\mathbf{2}d_4$ f_4 a_2	2	240	60480
c_7 $\mathbf{2}a_4$ f_4 a_2	1	120	60480
c_7 f_4^2 $\mathbf{2}a_2$	2	60	15120
e_7 $\mathbf{2}a_4$ f_4 b_2	1	110	55440
a_7 b_4^2 $\mathbf{2}a_1^2$	4	128	16128
a_7 f_4^2 $\mathbf{2}a_1^2$	4	160	20160
$\mathbf{2}a_7$ a_4 c_4 a_1^2	2	200	50400
a_7 $\mathbf{2}a_4$ c_3^2	2	240	60480
a_7 $\mathbf{2}d_4$ c_3^2	4	224	28224
a_7 f_4 $\mathbf{2}a_3$ c_3	2	192	48384
c_7 $\mathbf{2}d_4$ c_3^2	2	112	28224
c_7 f_4 $\mathbf{2}a_3$ c_3	1	88	44352
$\mathbf{2}a_7$ c_4 a_3 a_2 a_1	2	216	54432
a_7 b_4 $\mathbf{2}a_3$ b_2 $\mathbf{2}a_1$	2	160	40320
e_7 $\mathbf{2}a_4$ b_3 a_2 $\mathbf{2}a_1$	1	240	120960
e_7 f_4 b_3 $\mathbf{2}a_2$ $\mathbf{2}a_1$	1	120	60480
a_7 $\mathbf{2}a_3$ b_3 c_3 $\mathbf{2}a_1$	2	208	52416
a_7 $\mathbf{2}a_3$ c_3^2 $\mathbf{2}a_1$	4	288	36288
e_7 $\mathbf{2}a_3$ b_3 c_3 $\mathbf{2}a_1$	1	176	88704
a_7 $\mathbf{2}a_3^2$ b_2^2	4	192	24192
a_7 $\mathbf{2}a_3$ c_3 $\mathbf{2}a_2$ b_2	2	240	60480
d_7 $\mathbf{2}a_3^2$ b_2^2	4	176	22176
$\mathbf{2}a_7$ a_3^2 b_2 a_1^2	2	320	80640
$\mathbf{2}a_6$ e_6 c_5	1	126	63504
b_6 c_6 c_5	1	68	34272
c_6^2 b_5	2	92	23184

104   APPENDIX A

type of diagram	group	unit volume $[u=(16*16!)^{-1}]$	total volume $[u=1891890^{-1}]$
$c_6{}^2$ $2d_5$	2	88	22176
c_6 $2d_6$ c_5	1	64	32256
c_6 $2e_6$ a_5	2	138	34776
$2d_6$ e_6 c_5	1	108	54432
e_6 $2e_6$ c_5	2	102	25704
a_6 $2e_6$ f_4 a_1	1	105	52920
$2a_6$ c_6 a_4 a_1	1	210	105840
b_6 c_6 a_4 $2a_1$	1	140	70560
b_6 c_6 c_4 a_1	1	92	46368
b_6 c_6 f_4 a_1	1	50	25200
b_6 e_6 f_4 $2a_1$	1	60	30240
$c_6{}^2$ $2a_4$ a_1	2	180	45360
$c_6{}^2$ b_4 $2a_1$	2	120	30240
e_6 $2e_6$ a_4 $2a_1$	2	210	52920
e_6 $2e_6$ f_4 a_1	2	75	18900
a_6 b_6 c_3 $2a_2$	1	126	63504
b_6 c_6 a_3 $2a_2$	1	108	54432
b_6 c_6 b_3 a_2	1	78	39312
b_6 d_6 b_3 b_2	1	84	42336
e_6 $2e_6$ a_3 $2a_2$	2	162	40824
e_6 $2e_6$ b_3 a_2	2	117	29484
d_6 $2d_6$ c_3 a_1 $2a_1$	1	152	76608
a_6 $2e_6$ $a_2{}^2$ $2a_1$	2	378	95256
$c_6{}^2$ $2a_2$ b_2 $2a_1$	2	192	48384
c_6 a_5 $2a_5$ a_1	2	288	72576
c_6 a_5 b_5 $2a_1$	1	192	96768
c_6 $2a_5$ c_5 a_1	1	132	66528
d_6 $b_5{}^2$ a_1	2	84	21168
$2d_6$ $c_5{}^2$ a_1	2	84	21168
$2d_6$ c_5 d_5 $2a_1$	1	112	56448
e_6 $2a_5$ c_5 $2a_1$	2	156	39312
$2e_6$ $a_5{}^2$ $2a_1$	2	288	72576
a_6 $2d_5$ f_4 a_2	1	168	84672
b_6 d_5 f_4 b_2	1	64	32256
c_6 b_5 a_4 $2a_2$	1	150	75600
c_6 b_5 f_4 a_2	1	78	39312
c_6 c_5 b_4 $2a_2$	1	96	48384
d_6 b_5 b_4 a_2	1	132	66528
d_6 b_5 f_4 b_2	1	84	42336
$2d_6$ c_5 a_4 b_2	1	120	60480
e_6 $2a_5$ c_4 $2a_2$	2	126	31752
e_6 $2a_5$ f_4 b_2	2	84	21168
$2e_6$ a_5 a_4 $2a_2$	2	225	56700
$2e_6$ a_5 f_4 a_2	2	117	29484
b_6 a_5 c_4 a_1 $2a_1$	1	192	96768

type of diagram	group	unit volume $[u=(16*16!)^{-1}]$	total volume $[u=1891890^{-1}]$
d_6 $2a_5$ c_4 a_1 $2a_1$	1	216	108864
d_6 b_5 b_4 a_1^2	1	152	76608
d_6 b_5 c_4 a_1 $2a_1$	1	208	104832
a_6 $2a_5$ b_3 c_3	1	126	63504
b_6 a_5 b_3 c_3	1	96	48384
c_6 b_5 a_3 b_3	1	112	56448
d_6 $2a_5$ b_3 c_3	1	108	54432
d_6 b_5 b_3 c_3	1	104	52416
e_6 $2a_5$ b_3 c_3	2	102	25704
$2e_6$ a_5 a_3 b_3	2	168	42336
$2a_6$ a_5 a_3 b_2 a_1	1	336	169344
c_6 c_5 $2a_3$ b_2 $2a_1$	1	144	72576
$2d_6$ a_5 c_3 b_2 a_1	1	168	84672
d_6 b_5 a_3 $2a_2$ a_1	1	336	169344
d_6 b_5 $2a_3$ a_2 a_1	1	240	120960
$2d_6$ c_5 c_3 a_2 a_1	1	132	66528
c_6 a_5 $2a_2^2$ b_2	2	324	81648
$2e_6$ a_5 a_2^2 $2a_2$	2	297	74844
a_6 $2a_4$ f_4 c_3	1	140	70560
b_6 a_4 f_4 c_3	1	70	35280
c_6 b_4 c_4 b_3	1	76	38304
c_6 b_4 f_4 c_3	1	60	30240
c_6 f_4^2 b_3	2	46	11592
d_6 $2a_4$ c_4 b_3	1	140	70560
d_6 b_4^2 a_3	2	192	48384
d_6 b_4 c_4 b_3	1	136	68544
a_6 f_4^2 $2a_2$ $2a_1$	2	126	31752
b_6 a_4 c_4 $2a_2$ a_1	1	150	75600
b_6 c_4 f_4 a_2 a_1	1	78	39312
c_6 b_4^2 b_2 a_1	2	76	19152
c_6 b_4 c_4 b_2 $2a_1$	1	104	52416
d_6 b_4^2 b_2 a_1	2	136	34272
d_6 $2d_4^2$ a_1^3	4	240	30240
b_6 c_4 a_3 b_3 a_1	1	112	56448
d_6 c_4 b_3^2 $2a_1$	2	176	44352
d_6 f_4 b_3^2 $2a_1$	2	104	26208
e_6 f_4 b_3 $2a_2^2$	2	99	24948
c_6 a_4 $2a_3$ $2a_2$ b_2	1	240	120960
c_6 b_4 a_3 $2a_2$ b_2	1	168	84672
c_6 b_4 $2a_3$ a_2 b_2	1	120	60480
$2e_6$ a_4 b_3 a_2^2	2	225	56700
$2e_6$ f_4 a_3 a_2^2	2	162	40824
d_6 b_4 b_3 b_2 a_1 $2a_1$	1	176	88704
d_6 c_4 b_3 $2a_2$ a_1 $2a_1$	1	288	145152
d_6 $2d_4$ $2a_3$ a_2 a_1^2	2	384	96768

APPENDIX A

type of diagram	group	unit volume $[u=(16*16!)^{-1}]$	total volume $[u=1891890^{-1}]$
d_6 c_4 $2a_3$ a_1^2 $2a_1^2$	2	448	112896
d_6 $2d_4$ c_3 a_1^2 $2a_1^2$	2	320	80640
c_6 a_3 $2a_3^2$ b_2	2	176	44352
d_6 b_3^3 b_2	2	120	30240
d_6 b_3^2 c_3 $2a_2$	2	144	36288
d_6 a_3 $2a_3^2$ a_1^2	4	576	72576
d_6 $2a_3$ b_3 c_3 a_1 $2a_1$	1	224	112896
$2d_6$ a_3 c_3^2 a_1^2	2	192	48384
a_5^2 $2a_5$ b_2	2	360	90720
a_5^2 b_5 $2a_2$	2	324	81648
a_5 $2a_5$ c_5 b_2	2	180	45360
b_5^2 d_5 a_1^2	2	112	28224
a_5 $2a_5$ a_4 b_3	2	240	60480
a_5 $2a_5$ c_4 b_3	2	120	30240
a_5 $2a_5$ f_4 c_3	2	102	25704
a_5 b_5 a_4 $2a_3$	1	240	120960
a_5 b_5 b_4 a_3	1	168	84672
a_5 b_5 f_4 c_3	1	96	48384
b_5^2 a_4 a_3	2	120	30240
c_5^2 b_4 $2a_3$	2	72	18144
d_5^2 $2a_4$ $2a_3$	2	240	60480
d_5^2 $2d_4$ $2a_3$	4	224	28224
a_5^2 $2a_4$ $2a_2$ $2a_1$	2	360	90720
b_5^2 d_4 a_2 a_1	2	132	33264
b_5 d_5 $2a_4$ a_2 a_1	1	180	90720
b_5 d_5 c_4 $2a_2$ a_1	1	168	84672
c_5^2 $2d_4$ b_2 $2a_1$	2	96	24192
a_5 $2a_5$ c_4 a_1^2 $2a_1$	2	312	78624
a_5 $2a_5$ b_3 c_3 a_1	2	156	39312
a_5 $2a_5$ c_3^2 a_1	2	216	54432
b_5 d_5 a_3 $2a_3$ a_1	1	256	129024
d_5^2 $2a_3^2$ $2a_1$	8	288	18144
a_5^2 $2a_3$ $2a_2$ b_2	2	432	108864
a_5^2 b_3 $2a_2^2$	8	396	24948
a_5 $2a_5$ c_3 a_2 b_2	2	252	63504
$2a_5$ c_5 c_3 a_2^2	2	216	54432
a_5 $2a_5$ a_3 b_2 a_1 $2a_1$	2	384	96768
a_5 $2a_5$ b_3 a_2 a_1 $2a_1$	2	252	63504
a_5^2 $2a_2^3$ $2a_1$	8	432	27216
a_5^2 $2a_2^2$ b_2 $2a_1$	8	504	31752
a_5 a_4 $2a_4^2$	2	225	56700
b_5 a_4^2 $2a_4$	2	175	44100
c_5 b_4^2 c_4	2	52	13104
c_5 f_4^3	6	34	2856
d_5 a_4 $2a_4^2$	2	200	50400

type of diagram	group	unit volume $[u=(16*16!)^{-1}]$	total volume $[u=1891890^{-1}]$
$d_5\ b_4{}^2\ c_4$	2	96	24192
$d_5\ d_4\ \mathbf{2}d_4{}^2$	4	160	20160
$a_5\ b_4\ c_4\ b_3\ \mathbf{2}a_1$	1	144	72576
$a_5\ f_4{}^2\ b_3\ \mathbf{2}a_1$	2	96	24192
$b_5\ b_4\ d_4\ a_3\ a_1$	1	192	96768
$b_5\ c_4\ d_4\ b_3\ a_1$	1	136	68544
$b_5\ c_4\ f_4\ a_3\ a_1$	1	112	56448
$a_5\ f_4{}^2\ \mathbf{2}a_2{}^2$	4	99	12474
$\mathbf{2}a_5\ c_4\ f_4\ a_2{}^2$	2	126	31752
$a_5\ b_4{}^2\ b_2\ a_1\ \mathbf{2}a_1$	2	144	36288
$d_5\ f_4\ b_3{}^2\ \mathbf{2}a_2$	2	84	21168
$\mathbf{2}d_5\ f_4\ a_3{}^2\ a_2$	2	240	60480
$a_5\ \mathbf{2}d_4\ c_3{}^2\ a_1\ \mathbf{2}a_1$	2	240	60480
$c_5\ \mathbf{2}d_4\ c_3{}^2\ a_1\ \mathbf{2}a_1$	2	120	30240
$d_5\ c_4\ \mathbf{2}a_3{}^2\ a_1{}^2$	4	288	36288
$a_5\ c_4\ b_3\ \mathbf{2}a_2{}^2\ a_1$	2	252	63504
$\mathbf{2}a_5\ c_4\ a_3\ a_2{}^2\ \mathbf{2}a_1$	2	360	90720
$d_5\ b_3{}^4$	8	80	5040
$\mathbf{2}d_5\ a_3{}^2\ c_3{}^2$	4	224	28224
$a_5\ a_3\ \mathbf{2}a_3{}^2\ b_2\ a_1$	2	384	96768
$c_5\ \mathbf{2}a_3\ c_3{}^2\ a_2\ \mathbf{2}a_1$	2	192	48384
$d_5\ a_3\ \mathbf{2}a_3{}^2\ b_2\ a_1$	4	352	44352
$c_5\ c_3{}^3\ \mathbf{2}a_1{}^3$	6	160	13440
$a_5\ b_3\ a_2{}^2\ \mathbf{2}a_2{}^2\ \mathbf{2}a_1$	4	432	54432
$a_5\ a_2{}^2\ \mathbf{2}a_2{}^4$	8	405	25515
$a_4{}^2\ \mathbf{2}a_4{}^2\ a_1$	8	275	17325
$a_4\ b_4{}^2\ c_4\ \mathbf{2}a_1$	2	100	25200
$a_4\ f_4{}^3\ \mathbf{2}a_1$	6	70	5880
$b_4{}^2\ c_4{}^2\ a_1$	4	68	8568
$b_4{}^2\ d_4{}^2\ a_1$	4	240	30240
$b_4\ c_4{}^2\ f_4\ a_1$	2	76	19152
$c_4{}^4\ \mathbf{2}a_1$	24	176	3696
$c_4{}^3\ f_4\ \mathbf{2}a_1$	6	104	8736
$d_4{}^2\ \mathbf{2}d_4{}^2\ a_1$	24	208	4368
$f_4{}^4\ a_1$	24	25	525
$a_4\ f_4{}^2\ b_3\ \mathbf{2}a_2$	2	75	18900
$\mathbf{2}a_4\ c_4\ f_4\ a_3\ a_2$	1	180	90720
$b_4\ c_4{}^2\ b_3\ a_2$	2	108	27216
$b_4\ c_4{}^2\ c_3\ b_2$	2	88	22176
$c_4{}^3\ b_3\ b_2$	6	120	10080
$c_4{}^3\ c_3\ \mathbf{2}a_2$	6	144	12096
$c_4{}^2\ f_4\ a_3\ \mathbf{2}a_2$	2	168	42336
$c_4{}^2\ f_4\ \mathbf{2}a_3\ a_2$	2	120	30240
$f_4{}^3\ a_3\ \mathbf{2}a_2$	6	54	4536

APPENDIX A

type of diagram	group	unit volume $[u=(16*16!)^{-1}]$	total volume $[u=1891890^{-1}]$
$f_4^3\ b_3\ a_2$	6	39	3276
$b_4^2\ c_4\ a_2\ b_2\ a_1$	2	108	27216
$b_4\ c_4^2\ b_2^2\ a_1$	2	120	30240
$b_4\ d_4^2\ b_2^2\ \mathbf{2}a_1$	4	288	36288
$d_4^2\ \mathbf{2}d_4\ b_2\ \mathbf{2}a_1^3$	12	256	10752
$d_4\ \mathbf{2}d_4^2\ a_2\ a_1^3$	12	336	14112
$a_4\ \mathbf{2}d_4\ c_3^3$	6	160	13440
$c_4^2\ a_3\ b_3^2$	4	160	20160
$c_4^2\ \mathbf{2}a_3\ c_3^2$	4	112	14112
$c_4\ \mathbf{2}d_4\ c_3^3$	6	80	6720
$c_4\ f_4\ a_3^2\ \mathbf{2}a_3$	2	256	64512
$d_4\ f_4\ b_3^3$	6	68	5712
$\mathbf{2}d_4\ f_4\ a_3^3$	6	352	29568
$f_4^2\ a_3\ b_3^2$	4	56	7056
$b_4\ c_4\ a_3\ b_3\ b_2\ a_1$	1	160	80640
$c_4\ d_4\ b_3^2\ b_2\ a_1$	2	208	52416
$c_4\ d_4\ b_3^2\ a_1^2\ \mathbf{2}a_1$	2	288	72576
$b_4\ d_4\ a_3\ \mathbf{2}a_2\ b_2^2$	2	240	60480
$d_4^2\ b_3\ b_2^3$	12	352	14784
$b_4^2\ a_3\ b_2^2\ a_1^2$	4	160	20160
$b_4\ d_4\ b_3\ b_2^2\ a_1^2$	2	208	52416
$d_4^2\ \mathbf{2}a_3\ b_2^2\ \mathbf{2}a_1^2$	8	320	20160
$\mathbf{2}d_4^2\ a_3\ a_1^6$	24	512	10752
$d_4\ \mathbf{2}d_4\ c_3\ a_1^3\ \mathbf{2}a_1^3$	6	352	29568
$c_4^2\ b_2^4\ \mathbf{2}a_1$	8	144	9072
$d_4^2\ \mathbf{2}a_2\ b_2^3\ \mathbf{2}a_1$	12	384	16128
$d_4^2\ b_2^4\ \mathbf{2}a_1$	48	448	4704
$c_4\ a_3^2\ \mathbf{2}a_3^2\ a_1$	8	352	22176
$d_4\ b_3^4\ a_1$	24	104	2184
$d_4\ b_3^3\ c_3\ a_1$	6	144	12096
$\mathbf{2}d_4\ a_3^4\ \mathbf{2}a_1$	48	896	9408
$\mathbf{2}d_4\ c_3^4\ a_1$	24	104	2184
$a_4\ a_3\ \mathbf{2}a_3^2\ b_2^2$	4	240	30240
$b_4\ a_3^2\ \mathbf{2}a_3\ b_2^2$	4	192	24192
$d_4\ a_3\ \mathbf{2}a_3^2\ b_2^2$	8	224	14112
$d_4\ a_3\ b_3^2\ b_2^2$	4	256	32256
$d_4\ b_3^3\ a_2\ b_2$	6	168	14112
$c_4\ a_3^2\ \mathbf{2}a_3\ a_2\ \mathbf{2}a_1^2$	4	576	72576
$c_4\ a_3\ b_3^2\ \mathbf{2}a_2\ a_1^2$	2	240	60480
$\mathbf{2}d_4\ a_3^3\ \mathbf{2}a_2\ a_1^2$	12	768	32256
$\mathbf{2}d_4\ c_3^3\ a_2\ a_1\ \mathbf{2}a_1$	6	168	14112
$c_4\ a_3^3\ \mathbf{2}a_1^4$	24	896	18816
$c_4\ b_3^3\ a_1^4$	6	208	17472
$\mathbf{2}d_4\ a_3^2\ \mathbf{2}a_3\ a_1^4$	8	640	40320

type of diagram	group	unit volume $[u=(16*16!)^{-1}]$	total volume $[u=1891890^{-1}]$
$\mathbf{2}d_4\ a_3\ c_3{}^2\ a_1{}^2\ \mathbf{2}a_1{}^2$	4	256	32256
$d_4\ b_3{}^2\ b_2{}^2\ a_1{}^3$	4	352	44352
$d_4\ b_3\ b_2{}^3\ a_1{}^3\ \mathbf{2}a_1$	6	448	37632
$d_4\ c_3\ a_1{}^4\ \mathbf{2}a_1{}^6$	24	768	16128
$c_4\ b_2{}^6\ a_1$	24	176	3696
$d_4\ b_2{}^5\ a_1{}^3$	24	576	12096
$d_4\ a_1{}^5\ \mathbf{2}a_1{}^8$	192	896	2352
$a_3{}^2\ \mathbf{2}a_3{}^2\ c_3\ b_2$	8	224	14112
$a_3{}^2\ \mathbf{2}a_3\ c_3{}^2\ \mathbf{2}a_1{}^2$	8	320	20160
$a_3{}^4\ \mathbf{2}a_3\ \mathbf{2}a_1{}^2$	32	1024	16128
$a_3{}^3\ \mathbf{2}a_3{}^2\ a_1{}^2$	16	768	24192
$a_3{}^2\ \mathbf{2}a_3{}^2\ a_2\ b_2\ a_1$	8	480	30240
$a_3{}^2\ \mathbf{2}a_3{}^2\ b_2{}^2\ a_1$	16	288	9072
$a_3{}^4\ \mathbf{2}a_2\ \mathbf{2}a_1{}^3$	48	1152	12096
$a_3\ b_3{}^3\ b_2\ a_1{}^3$	6	256	21504
$b_3{}^4\ a_2\ a_1{}^3$	24	168	3528
$c_3{}^4\ b_2\ \mathbf{2}a_1{}^3$	24	128	2688
$a_3{}^4\ \mathbf{2}a_1{}^5$	192	1280	3360
$c_3{}^4\ a_1\ \mathbf{2}a_1{}^4$	24	176	3696
$a_3\ a_2{}^3\ \mathbf{2}a_2{}^4$	24	486	10206
$b_3\ a_2{}^4\ \mathbf{2}a_2{}^3$	24	405	8505
$c_3\ b_2{}^7$	168	112	336
$a_3\ b_2{}^6\ a_1{}^2$	48	384	4032
$b_3\ b_2{}^4\ a_1{}^6$	24	576	12096
$a_3\ a_1{}^6\ \mathbf{2}a_1{}^8$	384	1024	1344
$c_3\ a_1{}^7\ \mathbf{2}a_1{}^7$	168	896	2688
$a_2{}^4\ \mathbf{2}a_2{}^4\ a_1$	192	567	1488.375
$a_2\ b_2{}^7\ a_1$	168	240	720
$b_2{}^8\ a_1$	1344	144	54
$b_2{}^4\ a_1{}^9$	192	1792	4704
$b_2{}^4\ a_1{}^8\ \mathbf{2}a_1$	192	1280	3360
$a_2\ a_1{}^7\ \mathbf{2}a_1{}^8$	1344	1152	432
$b_2\ a_1{}^{15}$	20160	10240	256
$b_2\ a_1{}^8\ \mathbf{2}a_1{}^7$	1344	1024	384
$a_1{}^{16}\ \mathbf{2}a_1$	322560	12288	19.2
$a_1{}^9\ \mathbf{2}a_1{}^8$	10752	1280	60
			30270240

Total volume of 475 types (units of 1) = $30270240/1891890 = 16 = \sqrt{2}^8$.

Appendix B

There are 18 hyperbolic Lie algebras of rank greater or equal to 7, falling into 5 families, AE_n, BE_n, CE_n, DE_n, and $T_{r,s,t}$. We list all of them and give multiplicities for a limited number of roots of small height. For simply laced algebras, we compare to the global upper bound given in theorem 1.4. (Note that this theorem does not apply to the families BE_n and CE_n.) Even though we can only give very few values they still provide some indication of the quality of the upper bounds. It may be observed that the multiplicities for roots of level 1 all conform to formula (6.8). The values

$$p_{\text{rank}-2}(1 - \frac{r^2}{2})$$

provide material for many intriguing conjectures in this context. They are printed for this purpose only. From the point of disproving any existing conjectures the table of $T_{4,3,3}$ may be the most interesting. It shows that $T_{4,3,3}$ contains roots of multiplicities both larger and smaller than p_6. All root multiplicities in this appendix were calculated by a program based on the Peterson recursion formula ([**Kac90**], p.210).

The algebra $E_{10} = T_{7,3,2}$ has been studied extensively in [**KMW88**] where the root multiplicities for level 0,1, and 2 were determined explicitly. The table for E_{10} contains, for some small negative norms r^2, the explicit multiplicities as determined by [**KMW88**] for levels 1 and 2, and the global upper bound as provided by theorem 1.4. We see that this bound provides a reasonably good approximation.

APPENDIX B

$H_4^{(7)} = AE_7 \subset \mathcal{G}_3$

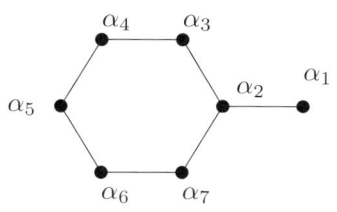

root-coefficients	norm	mult	thm 1.4	p_5
0, 1, 1, 1, 1, 1, 1	0	5	5	5
1, 2, 2, 2, 2, 2, 2	-2	20	21	20
2, 4, 3, 2, 1, 2, 3	-2	20	21	20
2, 4, 3, 2, 2, 2, 3	-4	65	71	65
1, 3, 3, 3, 3, 3, 3	-4	65	71	65
2, 4, 3, 3, 3, 3, 3	-6	189	217	190
2, 5, 4, 3, 2, 3, 4	-6	190	217	190
1, 4, 4, 4, 4, 4, 4	-6	190	217	190
3, 6, 5, 4, 3, 2, 4	-6	190	217	190
2, 5, 4, 3, 3, 3, 4	-8	502	603	506
2, 4, 4, 4, 4, 4, 4	-8	500	603	506

$H_4^{(8)} = AE_8 \subset \mathcal{G}_3$

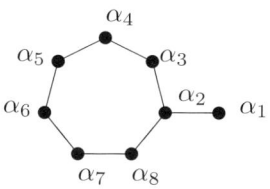

root-coefficients	norm	mult	thm 1.4	p_6
0, 1, 1, 1, 1, 1, 1, 1	0	6	6	6
1, 2, 2, 2, 2, 2, 2, 2	-2	27	28	27
2, 4, 3, 2, 1, 1, 2, 3	-2	27	28	27
2, 4, 3, 2, 2, 2, 2, 3	-4	97	105	98
1, 3, 3, 3, 3, 3, 3, 3	-4	98	105	98
2, 4, 3, 3, 3, 3, 3, 3	-6	310	350	315
2, 5, 4, 3, 2, 2, 3, 4	-6	309	350	315
3, 6, 5, 4, 3, 2, 2, 4	-6	309	350	315
3, 6, 4, 2, 2, 3, 4, 5	-6	309	350	315
1, 4, 4, 4, 4, 4, 4, 4	-6	315	350	315
2, 5, 4, 3, 3, 3, 3, 4	-8	894	1057	918

$H_4^{(9)} = AE_9 \subset \mathcal{G}_2$

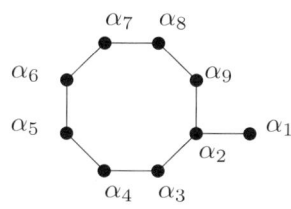

root-coefficients	norm	mult	thm 1.4	p_7
0, 1, 1, 1, 1, 1, 1, 1, 1	0	7	7	7
1, 2, 2, 2, 2, 2, 2, 2, 2	-2	35	36	35
2, 4, 3, 2, 1, 1, 1, 2, 3	-2	34	36	35
2, 4, 3, 2, 2, 2, 2, 2, 3	-4	136	148	140
1, 3, 3, 3, 3, 3, 3, 3, 3	-4	140	148	140
2, 5, 4, 3, 2, 1, 2, 3, 4	-4	132	148	140
3, 6, 5, 4, 3, 2, 1, 2, 4	-4	133	148	140
3, 6, 4, 2, 1, 2, 3, 4, 5	-4	133	148	140
2, 5, 4, 3, 2, 2, 2, 3, 4	-6	464	534	490
2, 4, 3, 3, 3, 3, 3, 3, 3	-6	475	534	490
2, 5, 4, 3, 3, 3, 3, 3, 4	-8	1464	1738	1547

APPENDIX B

$H_2^{(7)} = BE_7$

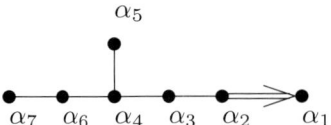

root-coefficients	norm	mult
2, 2, 2, 2, 1, 1, 0	0	5
3, 3, 3, 3, 1, 2, 1	-2	6
4, 4, 4, 4, 2, 2, 1	-4	21
5, 5, 5, 5, 2, 3, 1	-6	26
3, 4, 5, 6, 3, 4, 2	-6	26
4, 4, 5, 6, 3, 4, 2	-8	71
6, 6, 6, 6, 3, 3, 1	-8	71
6, 6, 6, 6, 2, 4, 2	-8	71
5, 5, 5, 6, 3, 4, 2	-10	91
7, 7, 7, 7, 3, 4, 1	-10	91
6, 6, 6, 6, 3, 4, 2	-12	217

$H_2^{(8)} = BE_8$

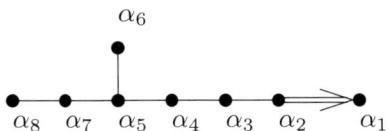

root-coefficients	norm	mult
2, 2, 2, 2, 2, 1, 1, 0	0	6
3, 3, 3, 3, 3, 1, 2, 1	-2	7
4, 4, 4, 4, 4, 2, 2, 1	-4	28
2, 3, 4, 5, 6, 3, 4, 2	-4	28
3, 3, 4, 5, 6, 3, 4, 2	-6	34
5, 5, 5, 5, 5, 2, 3, 1	-6	34
4, 4, 4, 5, 6, 3, 4, 2	-8	105
6, 6, 6, 6, 6, 3, 3, 1	-8	105
6, 6, 6, 6, 6, 2, 4, 2	-8	105
5, 5, 5, 5, 6, 3, 4, 2	-10	131
6, 6, 6, 6, 6, 3, 4, 2	-12	349

$H_2^{(9)} = BE_9$

root-coefficients	norm	mult
2, 2, 2, 2, 2, 2, 1, 1, 0	0	7
3, 3, 3, 3, 3, 3, 1, 2, 1	-2	8
1, 2, 3, 4, 5, 6, 3, 4, 2	-2	8
4, 4, 4, 4, 4, 4, 2, 2, 1	-4	36
2, 2, 3, 4, 5, 6, 3, 4, 2	-4	36
3, 3, 3, 4, 5, 6, 3, 4, 2	-6	42
5, 5, 5, 5, 5, 5, 2, 3, 1	-6	43
4, 4, 4, 4, 5, 6, 3, 4, 2	-8	147
6, 6, 6, 6, 6, 6, 3, 3, 1	-8	148
6, 6, 6, 6, 6, 6, 2, 4, 2	-8	148
5, 5, 5, 5, 5, 6, 3, 4, 2	-10	179

$H_2^{(10)} = BE_{10}$

root-coefficients	norm	mult
2, 2, 2, 2, 2, 2, 2, 1, 1, 0	0	8
3, 3, 3, 3, 3, 3, 3, 1, 2, 1	-2	9
1, 1, 2, 3, 4, 5, 6, 3, 4, 2	-2	8
4, 4, 4, 4, 4, 4, 4, 2, 2, 1	-4	45
2, 2, 2, 3, 4, 5, 6, 3, 4, 2	-4	44

APPENDIX B

$H_3^{(7)} = CE_7$

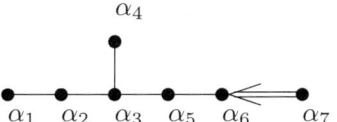

root-coefficients	norm	mult
0, 1, 2, 1, 2, 2, 1	0	4
0, 2, 4, 2, 4, 4, 2	0	5
1, 2, 4, 2, 4, 4, 2	-2	15
2, 4, 6, 3, 5, 4, 2	-4	44
1, 3, 6, 3, 6, 6, 3	-4	44
2, 4, 6, 2, 6, 6, 3	-4	50
2, 4, 6, 3, 6, 6, 3	-6	122
1, 4, 8, 4, 8, 8, 4	-6	121
2, 5, 8, 4, 7, 6, 3	-8	304
2, 5, 8, 3, 8, 8, 4	-8	304
2, 4, 8, 4, 8, 8, 4	-8	311
3, 6, 9, 4, 8, 7, 3	-10	725

$H_3^{(8)} = CE_8$

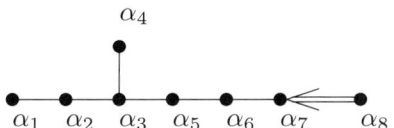

root-coefficients	norm	mult
0, 1, 2, 1, 2, 2, 2, 1	0	5
0, 2, 4, 2, 4, 4, 4, 2	0	6
1, 2, 4, 2, 4, 4, 4, 2	-2	21
2, 4, 6, 3, 5, 4, 3, 1	-2	21
2, 4, 6, 3, 5, 4, 4, 2	-4	71
1, 3, 6, 3, 6, 6, 6, 3	-4	70
2, 4, 6, 2, 6, 6, 6, 3	-4	77
2, 4, 6, 3, 6, 6, 6, 3	-6	215
2, 5, 8, 4, 7, 6, 5, 2	-6	218
1, 4, 8, 4, 8, 8, 8, 4	-6	212
2, 5, 8, 4, 7, 6, 6, 3	-8	596
3, 6, 9, 4, 8, 7, 6, 3	-10	1555

$H_3^{(9)} = CE_9$

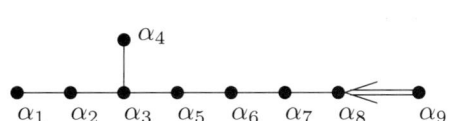

root-coefficients	norm	mult
0, 1, 2, 1, 2, 2, 2, 2, 1	0	6
0, 2, 4, 2, 4, 4, 4, 4, 2	0	7
1, 2, 4, 2, 4, 4, 4, 4, 2	-2	28
2, 4, 6, 3, 5, 4, 3, 2, 1	-2	29
2, 4, 6, 3, 5, 4, 4, 4, 2	-4	106
1, 3, 6, 3, 6, 6, 6, 6, 3	-4	104
2, 4, 6, 2, 6, 6, 6, 6, 3	-4	112
2, 4, 6, 3, 6, 6, 6, 6, 3	-6	349
2, 5, 8, 4, 7, 6, 5, 4, 2	-6	357
2, 5, 8, 4, 7, 6, 6, 6, 3	-8	1058

$H_3^{(10)} = CE_{10}$

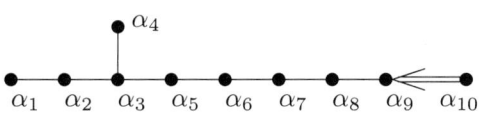

root-coefficients	norm	mult
0, 1, 2, 1, 2, 2, 2, 2, 2, 1	0	7
0, 2, 4, 2, 4, 4, 4, 4, 4, 2	0	8
1, 2, 4, 2, 4, 4, 4, 4, 4, 2	-2	36
2, 4, 6, 3, 5, 4, 3, 2, 2, 1	-2	37

$H_1^{(7)} = DE_7$

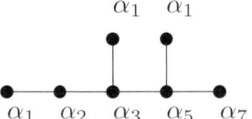

root-coefficients	norm	mult	thm 1.4	p_5
0, 1, 2, 1, 2, 1, 1	0	5	5	5
1, 2, 4, 2, 4, 2, 2	-2	20	21	20
2, 4, 6, 3, 5, 2, 2	-4	65	71	65
1, 3, 6, 3, 6, 3, 3	-4	65	71	65
2, 4, 6, 2, 6, 3, 3	-4	65	71	65
2, 4, 6, 3, 6, 3, 3	-6	190	217	190
1, 4, 8, 4, 8, 4, 4	-6	190	217	190
2, 5, 8, 4, 7, 3, 3	-8	506	603	506
2, 5, 8, 3, 8, 4, 4	-8	505	603	506
2, 4, 8, 4, 8, 4, 4	-8	505	603	506
2, 5, 8, 4, 8, 4, 4	-10	1263	1574	1265

$H_1^{(8)} = DE_8 \subset \mathcal{G}_3$

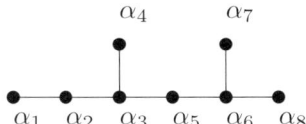

root-coefficients	norm	mult	thm 1.4	p_6
0, 1, 2, 1, 2, 2, 1, 1	0	6	6	6
1, 2, 4, 2, 4, 4, 2, 2	-2	27	28	27
2, 4, 6, 3, 5, 4, 2, 1	-2	27	28	27
2, 4, 6, 3, 5, 4, 1, 2	-2	27	28	27
2, 4, 6, 3, 5, 4, 2, 2	-4	98	105	98
1, 3, 6, 3, 6, 6, 3, 3	-4	98	105	98
2, 4, 6, 2, 6, 6, 3, 3	-4	98	105	98
2, 4, 6, 3, 6, 6, 3, 3	-6	314	350	315
2, 5, 8, 4, 7, 6, 3, 2	-6	315	350	315
2, 5, 8, 4, 7, 6, 2, 3	-6	315	350	315
2, 5, 8, 4, 7, 6, 3, 3	-8	914	1057	918

$H_1^{(9)} = DE_9$

root-coefficients	norm	mult	thm 1.4	p_7
0, 1, 2, 1, 2, 2, 2, 1, 1	0	7	7	7
1, 2, 4, 2, 4, 4, 4, 2, 2	-2	35	36	35
2, 4, 6, 3, 5, 4, 3, 1, 1	-2	35	36	35
2, 4, 6, 3, 5, 4, 4, 2, 2	-4	139	148	140
1, 3, 6, 3, 6, 6, 6, 3, 3	-4	140	148	140
2, 4, 6, 2, 6, 6, 6, 3, 3	-4	140	148	140
2, 4, 6, 3, 6, 6, 6, 3, 3	-6	485	534	490
2, 5, 8, 4, 7, 6, 5, 2, 2	-6	484	534	490
2, 5, 8, 4, 7, 6, 6, 3, 3	-8	1522	1738	1547

$H_1^{(10)} = DE_{10} \subset \mathcal{G}_2$

root-coefficients	norm	mult	thm 1.4	p_8
0, 1, 2, 1, 2, 2, 2, 2, 1, 1	0	8	8	8
1, 2, 4, 2, 4, 4, 4, 4, 2, 2	-2	44	45	44
2, 4, 6, 3, 5, 4, 3, 2, 1, 1	-2	43	45	44
1, 3, 6, 3, 6, 6, 6, 6, 3, 3	-4	192	201	192

APPENDIX B

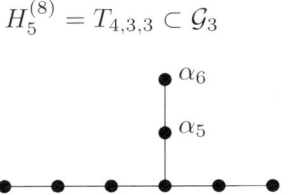

$H_5^{(8)} = T_{4,3,3} \subset \mathcal{G}_3$

root-coefficients	norm	mult	thm 1.4	p_6
0, 1, 2, 3, 2, 1, 2, 1	0	5	5	5
1, 2, 4, 6, 4, 2, 4, 2	-2	27	28	27
2, 4, 6, 8, 5, 2, 5, 2	-4	98	105	98
1, 3, 6, 9, 6, 3, 6, 3	-4	98	105	98
2, 4, 6, 9, 6, 3, 6, 3	-6	315	350	315
1, 4, 8, 12, 8, 4, 8, 4	-6	315	350	315
2, 5, 8, 11, 7, 3, 7, 3	-8	918	1057	918
2, 4, 8, 12, 8, 4, 8, 4	-8	917	1057	918
3, 6, 9, 12, 8, 4, 7, 2	-6	316	350	315
3, 6, 9, 12, 7, 2, 8, 4	-6	316	350	315
2, 5, 8, 12, 8, 4, 8, 4	-10	2491	2975	2492
3, 6, 9, 12, 8, 4, 7, 3	-10	2493	2975	2492
3, 6, 9, 12, 7, 3, 8, 4	-10	2493	2975	2492
3, 6, 9, 12, 8, 4, 8, 4	-12	6372	7883	6372
2, 6, 10, 14, 9, 4, 9, 4	-12	6368	7883	6372
3, 6, 10, 14, 9, 4, 9, 4	-14	15524	19900	15525

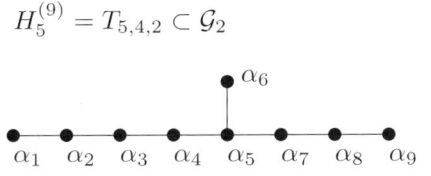

$H_5^{(9)} = T_{5,4,2} \subset \mathcal{G}_2$

root-coefficients	norm	mult	thm 1.4	p_7
0, 1, 2, 3, 4, 2, 3, 2, 1	0	7	7	7
1, 2, 4, 6, 8, 4, 6, 4, 2	-2	35	36	35
2, 4, 6, 8, 10, 5, 7, 4, 1	-2	35	36	35
2, 4, 6, 8, 10, 5, 7, 4, 2	-4	140	148	140
1, 3, 6, 9, 12, 6, 9, 6, 3	-4	140	148	140

$H_2^{(10)} = E_{10} = T_{7,3,2} \subset \mathcal{G}_2$

norm	level 0,1 [KMW88]	level 2 [KMW88]	thm 1.4
0	8	-	8
-2	44	44	45
-4	192	192	201
-6	726	727	780
-8	2464	2472	2718
-10	7704	7747	8730
-12	22528	22712	26226
-14	62337	63020	74556
-16	164560	166840	202180

Bibliography

[Apo76] Apostol, T.M.: *Modular Functions and Dirichlet Series in Number Theory*, Springer Verlag (Graduate Texts in Mathematics 41), New York, Berlin, Heidelberg, 1976

[BM79] Berman, S., Moody, R.V.: *Multiplicities in Lie Algebras*, Proc. Amer. Math. Soc. 76 (1979), 223–228

[Bor85] Borcherds, R.E.: *The Leech Lattice*, Proc. R. Soc. Lond. A 398 (1985), 365–376

[Bor86] Borcherds, R.E.: *Vertex Algebras, Kac-Moody Algebras, and the Monster*, Proc. Natl. Acad. Sci. USA 83 (1986), 3068-3071

[Bor88] Borcherds, R.E.: *Generalized Kac-Moody algebras*, J. Algebra 115 (1988), 501-512

[Bor90a] Borcherds, R.E.: *Lattices like the Leech Lattice*, J. Algebra 130 (1990), 219-234

[Bor90b] Borcherds, R.E.: *The Monster Lie algebra*, Adv. Math. 83 (1990), 30-47

[Bor91] Borcherds, R.E.: *Central extensions of generalized Kac-Moody algebras*, J. Algebra 140 (1991), 330-335

[Bor92] Borcherds, R.E.: *Monstrous Moonshine and Monstrous Lie Superalgebras*, Invent. Math. 109 (1992), 405-444

[CE56] Cartan, H., Eilenberg, S.: *Homological Algebra*, Princeton: Princeton University Press 1956

[Con85] Conway, J.H., Curtis, R.T., Norton, S.P., Parker, R.A., Wilson, R.A.: *ATLAS of Finite Groups*, Oxford Univ. Press, 1985

[CS83] Conway, J.H., Sloane, N.J.A.: *The Coxeter-Todd Lattice, the Mitchell Group, and Related Sphere Packings*, Proc. Camb. Phil. Soc. 93 (1983), 421-440

[CS88] Conway, J.H., Sloane, N.J.A.: *Sphere Packings, Lattices and Groups*, Springer Verlag (Grundlehren d. math. Wiss. 290), New York, Berlin, Heidelberg, 1988

[FF83] Feingold, A.J., Frenkel, I.B.: *A Hyperbolic Kac-Moody Algebra and the Theory of Siegel Modular Forms of Genus 2*, Math. Ann. 263 (1983), 87-144

[FLM88] Frenkel, I.B., Lepowsky, J., and Meurman, A.: *Vertex Operator Algebras and the Monster*, MA Academic Press, Boston, 1988

[GL76] Garland, H., Lepowsky, J.: *Lie algebra homology and the Macdonald-Kac formulas*, Invent. Math. 34 (1976), 37-76

[GN97] Gebert, R.W., Nicolai, H.: *On the imaginary simple roots of the Borcherds algebra* $\mathfrak{g}_{II_{9,1}}$, preprint IASSNS-HEP-97-53, AEI-937, 1997

[Jac85] Jacobson, N.: *Basic Algebra I*, 2nd ed., Freeman, New York, 1985

[Jan95] Jansen, C., Lux, K., Parker, R., Wilson, R.: *An Atlas of Brauer Characters*, Oxford University Press, 1995

[Jur96] Jurisich, E.: *An Exposition of Generalized Kac-Moody algebras*, Contemporary Maths 194 (1996)

[Jur98] Jurisich, E.: *Generalized Kac-Moody Lie algebras, free Lie algebras and the structure of the Monster Lie algebra*, J. Pure Appl. Algebra 126 (1998), 233-266

[Kac90] Kac, V.G.: *Infinite Dimensional Lie Algebras*, 3rd ed., Cambridge University Press, 1990

[KP83] Kac, V.G., Peterson, D.H.: *Regular Functions on Certain Infinite-dimensional Groups*, in: Arithmetic and Geometry, 141-166, Progress in Math. 36, Birkhäuser, Boston, 1983

[KMW88] Kac, V.G., Moody, R.V., and Wakimoto, M.: *On E_{10}*, Differential Methods in Theoretical Physics, Bleuler, K., Werner, M. (eds.), Kluwer Academic Publishers, 109-128

[Kob84] Koblitz, N.: *Introduction to Elliptic Curves and Modular Forms*, Springer Verlag (Graduate Texts in Mathematics), New York, Berlin, Heidelberg, 1984

[Rad29] Rademacher, H.: *Über die Erzeugenden von Kongruenzuntergruppen der Modulgruppe*, Abh. Math. Sem. Hamburg 7 (1929), 134-148

[Shi71] Shimura, G.: *Introduction to the Arithmetic Theory of Automorphic Functions*, Publ. Math. Soc. Japan 11, 1971

[Wan91] Wan Zhe-xian: *Introduction to Kac-Moody Algebra*, World Scientific Publishing, Singapore, 1991

Notation

The following list comprises all non-standard notations used in this work. It refers to the chapter where the notation is introduced and also gives a brief indication of its meaning. Note that we use the standard set in [**Kac90**] for finite and affine Lie algebras. They are therefore not listed below.

AE_n	B	hyperbolic Lie algebra
(A, j)	2.1	element of the metaplectic group
$\mathrm{Aut}(\Lambda^\sigma)$	5.3	group of automorphisms of Λ^σ that fix the origin
$\mathrm{Aut}(H)$	5.3	group of automorphisms fixing a hole H of $\mathcal{R}$
BE_n	B	hyperbolic Lie algebra
C	1.1.1, 5.2	Cartan matrix
$\mathcal{C}$	4.1	24-dimensional Golay-code
CE_n	B	hyperbolic Lie algebra
D	5.2	diagonal matrix
DE_n	B	hyperbolic Lie algebra
d	5.2	factor relating $\delta^\vee = \frac{1}{d}\nu^{-1}\delta$
E_r	1.5	part of M_Λ of grade $r \in II_{25,1}$
E_{10}	B	hyperbolic Lie algebra
e^r	1.3	element of the central extension of the Leech lattice
F	2.1	the Fricke involution
$G(C)$	1.1.1	GKM generated from generalized Cartan matrix C
$G^e(C)$	1.1.2	GKM centrally extended by degree derivations
$\mathcal{G}_N$	1.6	the GKM constructed in theorem 1.6
$GO_M^\epsilon(N)$	4.1	the general orthogonal group over $(\mathbb{Z}_N)^M$ of Witt type ϵ
H	5.2	subset of $\mathcal{R}$ that constitutes the vertices of a hole
$\langle H \rangle$	5.2	hole in its spatial meaning
$\mathcal{H}$	2.1	the upper half complex plane
$II_{n,1}$	1.4	$(n+1)$-dimensional even unimodular Lorentzian lattice
K_{12}	4.0	the Coxeter-Todd lattice
L	4.1	$\Lambda^{\sigma\perp}$
L	1.6, 5.1	$\Lambda^\sigma \oplus II_{1,1}$
L^*	3.1	the dual lattice of L
$L^\perp$	3.1	the orthogonal complement of L with respect to Λ
L^{*+}	1.5	the positive roots in L^*
L_n	1.3	operators forming the Virasoro algebra
M	1.6	$=\frac{24}{N+1}$
Mp	2.1	the metaplectic group, double cover of Γ
M_Λ	1.4	the fake monster Lie algebra
N	1.6	order of the automorphism σ, that is 2, 3, 5, 7, 11, or 23
$N\Delta$	5.3	Dynkin diagram entirely of long roots

NOTATION

$O_M^\epsilon(N)$	4.1	the 'generically simple' orthogonal group over $(\mathbb{Z}_N)^M$ of Witt type ϵ
P^n	1.3	physical space as subspace of a vertex algebra
p_σ	1.7	coefficients of the q expansion of $1/\eta_\sigma$
$\mathbb{Q}$		the rational numbers
Q	1.3	vertex operator
q	2.1	element of the unit disc, $q = e^{2\pi i \tau}$
$\mathbb{R}$		the real numbers
$\mathcal{R}$	5.1	elements of Λ^σ and $\Lambda^{\sigma*}$ representing the simple roots of $\mathcal{G}_N$
$\mathcal{R}_{dual}$	5.1	$= \mathcal{R} \cap \Lambda^{\sigma*}$
$\mathcal{R}_{fix}$	5.1	$= \mathcal{R} \cap \Lambda^\sigma$
$r_\mathcal{R}$	5.2	radius function
(r, L)	1.5, 5.1	greatest common divisor of $(r,v), v \in L$
S	2.1	one of the generators of Γ
S_n	1.4	the part of the symmetric algebra S of $\mathbb{Z}$ grading n
$SO_M^\epsilon(N)$	4.1	the special orthogonal group over $(\mathbb{Z}_N)^M$ of Witt type ϵ
$T_{r,s,t}$	B	hyperbolic Lie algebra
$U(C)$	1.1.4	universal GKM generated from generalized Cartan matrix C
$V(L)$	1.3	the vertex algebra of the lattice L
$(V_k, *)$	2.1	V_k is the typical generator of $\Gamma_0(N)$, $*$ stands for the unspecified branch of j (see (A,j))
W^σ	1.5	the Weyl group of the twisted algebra
$\mathbb{Z}$		the integers
$\mathbb{Z}_N$	4.1	the finite field $\mathbb{Z}/N\mathbb{Z}$
Γ	2.1	the modular group
$\Gamma_0(N)$	2.1	subgroup of Γ
Δ	5.3	arbitrary Dynkin diagram
$\Delta(H)$	5.2	Dynkin diagram associated to hole H
$\delta, \delta^\vee$	5.2	unique norm 0 vector of affine Lie algebra
η	2.2	the Dedekind eta-function
Θ, Θ_r	4.4	modular forms, right hand side of eqn. (1.25)
θ	3.1	the theta-function of a lattice
Λ	1.4	the Leech lattice
$\bigwedge$	1.2	the wedge product
Λ_{16}	4.0	the Barnes-Wall lattice
(λ, m, n)	1.4	typical element of $II_{25,1}$
ν	5.3	the isomorphism between Cartan subalgebra and its dual
$\pi_L, \pi_{\Lambda^\sigma}$	4.1	projections to the span of the relevant lattice
π_1, π_2	4.3	$\pi_L = \pi_1 \circ \pi_2$
$\rho, \rho^\vee$	1.1.3, 5.2	the Weyl vector of a Lie algebra
$\rho_M, \tilde{\rho}_M$	4.2	number of representations as sum of squares
σ	1.5	automorphism of the Leech lattice, of prime order
σ	1.6	automorphism of the Leech lattice, cycle shape $1^M N^M$
$\phi_{\sigma,V}$	1.7	$= \sum \text{Tr}(\sigma, V_n) q^n$
ψ_j	2.2	modular form, $\psi_j(\tau) = \eta(\frac{\tau+j}{N} + j)$
$[\cdot]$		space spanned by any basis
$(\cdot, \cdot)$	1.1.3	bilinear form

Editorial Information

To be published in the *Memoirs*, a paper must be correct, new, nontrivial, and significant. Further, it must be well written and of interest to a substantial number of mathematicians. Piecemeal results, such as an inconclusive step toward an unproved major theorem or a minor variation on a known result, are in general not acceptable for publication. Papers appearing in *Memoirs* are generally longer than those appearing in *Transactions*, which shares the same editorial committee.

As of January 31, 2002, the backlog for this journal was approximately 5 volumes. This estimate is the result of dividing the number of manuscripts for this journal in the Providence office that have not yet gone to the printer on the above date by the average number of monographs per volume over the previous twelve months, reduced by the number of volumes published in four months (the time necessary for preparing a volume for the printer). (There are 6 volumes per year, each containing at least 4 numbers.)

A Consent to Publish and Copyright Agreement is required before a paper will be published in the *Memoirs*. After a paper is accepted for publication, the Providence office will send a Consent to Publish and Copyright Agreement to all authors of the paper. By submitting a paper to the *Memoirs*, authors certify that the results have not been submitted to nor are they under consideration for publication by another journal, conference proceedings, or similar publication.

Information for Authors

Memoirs are printed from camera copy fully prepared by the author. This means that the finished book will look exactly like the copy submitted.

The paper must contain a *descriptive title* and an *abstract* that summarizes the article in language suitable for workers in the general field (algebra, analysis, etc.). The *descriptive title* should be short, but informative; useless or vague phrases such as "some remarks about" or "concerning" should be avoided. The *abstract* should be at least one complete sentence, and at most 300 words. Included with the footnotes to the paper should be the 2000 *Mathematics Subject Classification* representing the primary and secondary subjects of the article. The classifications are accessible from www.ams.org/msc/. The list of classifications is also available in print starting with the 1999 annual index of *Mathematical Reviews*. The Mathematics Subject Classification footnote may be followed by a list of *key words and phrases* describing the subject matter of the article and taken from it. Journal abbreviations used in bibliographies are listed in the latest *Mathematical Reviews* annual index. The series abbreviations are also accessible from www.ams.org/publications/. To help in preparing and verifying references, the AMS offers MR Lookup, a Reference Tool for Linking, at www.ams.org/mrlookup/. When the manuscript is submitted, authors should supply the editor with electronic addresses if available. These will be printed after the postal address at the end of the article.

Electronically prepared manuscripts. The AMS encourages electronically prepared manuscripts, with a strong preference for $\mathcal{AMS}$-LaTeX. To this end, the Society has prepared $\mathcal{AMS}$-LaTeX author packages for each AMS publication. Author packages include instructions for preparing electronic manuscripts, the *AMS Author Handbook*, samples, and a style file that generates the particular design specifications of that publication series. Though $\mathcal{AMS}$-LaTeX is the highly preferred format of TeX, author packages are also available in $\mathcal{AMS}$-TeX.

Authors may retrieve an author package from e-MATH starting from `www.ams.org/tex/` or via FTP to `ftp.ams.org` (login as `anonymous`, enter username as password, and type `cd pub/author-info`). The *AMS Author Handbook* and the *Instruction Manual* are available in PDF format following the author packages link from `www.ams.org/tex/`. The author package can be obtained free of charge by sending email to `pub@ams.org` (Internet) or from the Publication Division, American Mathematical Society, P.O. Box 6248, Providence, RI 02940-6248. When requesting an author package, please specify $\mathcal{AMS}$-LaTeX or $\mathcal{AMS}$-TeX, Macintosh or IBM (3.5) format, and the publication in which your paper will appear. Please be sure to include your complete mailing address.

Sending electronic files. After acceptance, the source file(s) should be sent to the Providence office (this includes any TeX source file, any graphics files, and the DVI or PostScript file).

Before sending the source file, be sure you have proofread your paper carefully. The files you send must be the EXACT files used to generate the proof copy that was accepted for publication. For all publications, authors are required to send a printed copy of their paper, which exactly matches the copy approved for publication, along with any graphics that will appear in the paper.

TeX files may be submitted by email, FTP, or on diskette. The DVI file(s) and PostScript files should be submitted only by FTP or on diskette unless they are encoded properly to submit through email. (DVI files are binary and PostScript files tend to be very large.)

Electronically prepared manuscripts can be sent via email to `pub-submit@ams.org` (Internet). The subject line of the message should include the publication code to identify it as a Memoir. TeX source files, DVI files, and PostScript files can be transferred over the Internet by FTP to the Internet node `e-math.ams.org` (130.44.1.100).

Electronic graphics. Comprehensive instructions on preparing graphics are available at `www.ams.org/jourhtml/graphics.html`. A few of the major requirements are given here.

Submit files for graphics as EPS (Encapsulated PostScript) files. This includes graphics originated via a graphics application as well as scanned photographs or other computer-generated images. If this is not possible, TIFF files are acceptable as long as they can be opened in Adobe Photoshop or Illustrator. No matter what method was used to produce the graphic, it is necessary to provide a paper copy to the AMS.

Authors using graphics packages for the creation of electronic art should also avoid the use of any lines thinner than 0.5 points in width. Many graphics packages allow the user to specify a "hairline" for a very thin line. Hairlines often look acceptable when proofed on a typical laser printer. However, when produced on a high-resolution laser imagesetter, hairlines become nearly invisible and will be lost entirely in the final printing process.

Screens should be set to values between 15% and 85%. Screens which fall outside of this range are too light or too dark to print correctly. Variations of screens within a graphic should be no less than 10%.

Inquiries. Any inquiries concerning a paper that has been accepted for publication should be sent directly to the Electronic Prepress Department, American Mathematical Society, P. O. Box 6248, Providence, RI 02940-6248.

Editors

This journal is designed particularly for long research papers, normally at least 80 pages in length, and groups of cognate papers in pure and applied mathematics. Papers intended for publication in the *Memoirs* should be addressed to one of the following editors. In principle the Memoirs welcomes electronic submissions, and some of the editors, those whose names appear below with an asterisk (*), have indicated that they prefer them. However, editors reserve the right to request hard copies after papers have been submitted electronically. Authors are advised to make preliminary email inquiries to editors about whether they are likely to be able to handle submissions in a particular electronic form.

Algebra to KAREN E. SMITH, Department of Mathematics, University of Michigan, 525 University, Suite 2832, Ann Arbor, MI 48109-1109; email: `kesmith@lsa.umich.edu`

Algebraic geometry and commutative algebra to LAWRENCE EIN, Department of Mathematics, University of Illinois, 851 S. Morgan (M/C 249), Chicago, IL 60607-7045; email: `ein@uic.edu`

Algebraic topology and cohomology of groups to STEWART PRIDDY, Department of Mathematics, Northwestern University, 2033 Sheridan Road, Evanston, IL 60208-2730; email: `priddy@math.nwu.edu`

Combinatorics and Lie theory to SERGEY FOMIN, Department of Mathematics, University of Michigan, Ann Arbor, Michigan 48109-1109; email: `fomin@math.lsa.umich.edu`

Complex analysis and complex geometry to DUONG H. PHONG, Department of Mathematics, Columbia University, 2990 Broadway, New York, NY 10027-0029; email: `phong@math.columbia.edu`

*__Differential geometry and global analysis__ to LISA C. JEFFREY, Department of Mathematics, University of Toronto, 100 St. George St., Toronto, ON Canada M5S 3G3; email: `jeffrey@math.toronto.edu`

Dynamical systems and ergodic theory to ROBERT F. WILLIAMS, Department of Mathematics, University of Texas, Austin, Texas 78712-1082; email: `bob@math.utexas.edu`

Functional analysis and operator algebras to DAN VOICULESCU, Department of Mathematics, University of California, Berkeley, 970 Evans Hall, Floor 9, Berkeley, CA 94720-0001; email: `dvv@math.berkeley.edu`

Geometric topology, knot theory and hyperbolic geometry to ABIGAIL A. THOMPSON, Department of Mathematics, University of California, Davis, Davis, CA 95616-5224; email: `thompson@math.ucdavis.edu`

Harmonic analysis, representation theory, and Lie theory to ROBERT J. STANTON, Department of Mathematics, The Ohio State University, 231 West 18th Avenue, Columbus, OH 43210-1174; email: `stanton@math.ohio-state.edu`

*__Logic__ to THEODORE SLAMAN, Department of Mathematics, University of California, Berkeley, CA 94720-3840; email: `slaman@math.berkeley.edu`

Number theory to HAROLD G. DIAMOND, Department of Mathematics, University of Illinois, 1409 W. Green St., Urbana, IL 61801-2917; email: `diamond@math.uiuc.edu`

*__Ordinary differential equations, partial differential equations, and applied mathematics__ to PETER W. BATES, Department of Mathematics, Michigan State University, East Lansing, MI 48824-1027; email: `bates@math.msu.edu`

*__Probability and statistics__ to KRZYSZTOF BURDZY, Department of Mathematics, University of Washington, Box 354350, Seattle, Washington 98195-4350; email: `burdzy@math.washington.edu`

*__Real and harmonic analysis and geometric partial differential equations__ to WILLIAM BECKNER, Department of Mathematics, University of Texas, Austin, TX 78712-1082; email: `beckner@math.utexas.edu`

All other communications to the editors should be addressed to the Managing Editor, WILLIAM BECKNER, Department of Mathematics, University of Texas, Austin, TX 78712-1082; email: `beckner@math.utexas.edu`.

Selected Titles in This Series

(*Continued from the front of this publication*)

716 **John H. Palmieri,** Stable homotopy over the Steenrod algebra, 2001
715 **W. N. Everitt and L. Markus,** Multi-interval linear ordinary boundary value problems and complex symplectic algebra, 2001
714 **Earl Berkson, Jean Bourgain, and Aleksander Pełczynski,** Canonical Sobolev projections of weak type $(1,1)$, 2001
713 **Dorina Mitrea, Marius Mitrea, and Michael Taylor,** Layer potentials, the Hodge Laplacian, and global boundary problems in nonsmooth Riemannian manifolds, 2001
712 **Raúl E. Curto and Woo Young Lee,** Joint hyponormality of Toeplitz pairs, 2001
711 **V. G. Kac, C. Martinez, and E. Zelmanov,** Graded simple Jordan superalgebras of growth one, 2001
710 **Brian Marcus and Selim Tuncel,** Resolving Markov chains onto Bernoulli shifts via positive polynomials, 2001
709 **B. V. Rajarama Bhat,** Cocylces of CCR flows, 2001
708 **William M. Kantor and Ákos Seress,** Black box classical groups, 2001
707 **Henning Krause,** The spectrum of a module category, 2001
706 **Jonathan Brundan, Richard Dipper, and Alexander Kleshchev,** Quantum Linear groups and representations of $GL_n(\mathbb{F}_q)$, 2001
705 **I. Moerdijk and J. J. C. Vermeulen,** Proper maps of toposes, 2000
704 **Jeff Hooper, Victor Snaith, and Min van Tran,** The second Chinburg conjecture for quaternion fields, 2000
703 **Erik Guentner, Nigel Higson, and Jody Trout,** Equivariant E-theory for C^*-algebras, 2000
702 **Ilijas Farah,** Analytic guotients: Theory of liftings for quotients over analytic ideals on the integers, 2000
701 **Paul Selick and Jie Wu,** On natural coalgebra decompositions of tensor algebras and loop suspensions, 2000
700 **Vicente Cortés,** A new construction of homogeneous quaternionic manifolds and related geometric structures, 2000
699 **Alexander Fel'shtyn,** Dynamical zeta functions, Nielsen theory and Reidemeister torsion, 2000
698 **Andrew R. Kustin,** Complexes associated to two vectors and a rectangular matrix, 2000
697 **Deguang Han and David R. Larson,** Frames, bases and group representations, 2000
696 **Donald J. Estep, Mats G. Larson, and Roy D. Williams,** Estimating the error of numerical solutions of systems of reaction-diffusion equations, 2000
695 **Vitaly Bergelson and Randall McCutcheon,** An ergodic IP polynomial Szemerédi theorem, 2000
694 **Alberto Bressan, Graziano Crasta, and Benedetto Piccoli,** Well-posedness of the Cauchy problem for $n \times n$ systems of conservation laws, 2000
693 **Doug Pickrell,** Invariant measures for unitary groups associated to Kac-Moody Lie algebras, 2000
692 **Mara D. Neusel,** Inverse invariant theory and Steenrod operations, 2000
691 **Bruce Hughes and Stratos Prassidis,** Control and relaxation over the circle, 2000
690 **Robert Rumely, Chi Fong Lau, and Robert Varley,** Existence of the sectional capacity, 2000
689 **M. A. Dickmann and F. Miraglia,** Special groups: Boolean-theoretic methods in the theory of quadratic forms, 2000

For a complete list of titles in this series, visit the
AMS Bookstore at **www.ams.org/bookstore/**.